Samiro Khodayar Pardo

High-resolution analysis of the initiation of deep convection forced by boundary-layer processes

Wissenschaftliche Berichte des Instituts für Meteorologie und Klimaforschung des Karlsruher Instituts für Technologie (KIT)
Band 50

Herausgeber: Prof. Dr. Ch. Kottmeier

Institut für Meteorologie und Klimaforschung
am Karlsruher Institut für Technologie (KIT)
Kaiserstr. 12, 76128 Karlsruhe

Eine Übersicht aller bisher in dieser Schriftenreihe
erschienenen Bände finden Sie am Ende des Buches.

High-resolution analysis of the initiation of deep convection forced by boundary-layer processes

by
Samiro Khodayar Pardo

KIT Scientific Publishing

Dissertation, Karlsruher Institut für Technologie (KIT)
Fakultät für Physik, 2009
Referenten: Prof. Dr. Christoph Kottmeier, Prof. Dr. Sarah Jones

Impressum

KIT Scientific Publishing

Karlsruher Institut für Technologie (KIT)
KIT Scientific Publishing
Straße am Forum 2
D-76131 Karlsruhe

KIT Scientific Publishing is a registered trademark of Karlsruhe Institute of Technology. Reprint using the book cover is not allowed.

www.ksp.kit.edu

Print on Demand 2015

ISSN 0179-5619
ISBN 978-3-86644-770-7
DOI: 10.5445/KSP/1000024934

High-resolution analysis of the initiation of deep convection forced by boundary-layer processes

Zur Erlangung des akademischen Grades eines

DOKTORS DER NATURWISSENSCHAFTEN

der Fakultät für Physik der Universität (TH) Karlsruhe

genehmigte

DISSERTATION

von

Dipl.-Phys. Samiro Khodayar Pardo

aus

Teheran (Iran)

Tag der mündlichen Prüfung: 12.06.2009
Referent: Prof. Dr. Christoph Kottmeier
Korreferent: Prof. Dr. Sarah Jones

Zusammenfassung

Unter Nutzung der Synergie verschiedener Instrumente wird die Auslösung atmosphärischer Konvektion untersucht. Der Einfluss einer höheren räumlichen Datenauflösung auf die Erfassung der Auslösung hochreichender Konvektion wird analysiert und es wird eine Methodik zur Bestimmung der Wahrscheinlichkeit für das Auftreten hochreichender Konvektion über ebenem und komplexem Gelände entwickelt. Hierzu werden Intensivmessphasen (IOPs - Intensive Observation Periods) des in Südengland durchgeführten CSIP (Convective Storm Initiation Project) und der in Südwestdeutschland und im Osten Frankreichs durchgeführten Messkampagne COPS (Convective and Orographically-Induced Precipitation Study) herangezogen.

Die atmosphärische Grenzschicht spielt eine entscheidende Rolle bei der Auslösung konvektiver Systeme, denn Konvektion ist häufig mit Prozessen verbunden, die durch Grenzschichtinhomogenitäten verursacht werden. Zum besseren Verständnis der Grenzschichtprozesse werden deshalb einige Wetterlagen ausgewählt, die durch nicht vorhandenen oder geringen synoptischen Antrieb gekennzeichnet sind. Die Wahrscheinlichkeit von Konvektion wird mit Hilfe von Konvektionsindizes, die die atmosphärische Stabilität und die verfügbare konvektiven Energie beschreiben, sowie von Mechanismen, die die Entstehung von Konvektion fördern oder hemmen, analysiert. Hochauflösende Flugzeugmessungen geben Hinweise auf die Empfindlichkeit der Konvektionsindizes gegenüber Variabilitäten der Grenzschichttemperatur- und feuchte. Es wird gezeigt, dass bei Verwendung der entlang einer Flugbahn von ca. 36 km Länge gemessenen maximalen und minimalen Temperatur- und Feuchtewerte ein Unterschied von 680 J kg^{-1} in der Labilitätsenergie (CAPE - Convective Available Potential Energy) bzw. -150 J kg^{-1} in der Konvektionshemmung (CIN - Convective Inhibition) besteht. D.h., die Wahrscheinlichkeit für Konvektion variiert innerhalb von 36 km deutlich. Daraus folgt, dass die für Konvektion relevante räumliche Variabilität des Temperatur- und insbesondere des Wasserdampffeldes in der Grenzschicht während der CSIP- und COPS-Kampagnen trotz des dichten Netzwerks von Radiosondenstationen im Abstand von 25 bis 50 km nicht ausreichend genug erfasst wurde.

Die räumliche Auflösung konnte durch Nutzung des synergetischen Effekts von Daten des Radiosondenmessnetzes, der automatischen Wetterstationen (AWS), der synoptischen Stationen (SYNOP) und insbesondere der GPS-Systeme erhöht werden. Da die GPS-Messungen den Säulenwasserdampfgehalt bestimmen, wird ein neuartiges Verfahren (Adjustment Method) vorgestellt, mit Hilfe dessen der integrierte Wasserdampf (IWV - Integrated Water Vapour) aus den GPS-Messungen mit den Radiosondenprofilen verknüpft werden kann. Die auf diese Weise kombinierten Datensätze erlauben die Berechnung der Konvektionsindizes mit einer für die Konvektionsanalyse notwendigen räumlichen und zeitlichen Auflösung. Durch die erhöhte räumliche Auflösung und die Kombination der Konvektionsindizes - basierend auf Daten von verschiedenen Wetterlagen und sowohl über ebenem (CSIP) als auch über komplexes Gelände (COPS) - konnten die Bedingungen erfolgreich ermittelt werden, bei denen eine hohe Wahrscheinlichkeit für hochreichende Konvektion besteht. Des Weiteren wird die räumliche und zeitliche Entwicklung stabiler Schichten oder Lids in der unteren oder mittleren Atmosphäre und deren Rolle bei der Unterdrückung von Konvektion untersucht. Höhe, Stärke und Fortdauer dieser stabilen Schichten zeigen sich dabei als für die Auslösung von Konvektion bedeutende Faktoren. Stabile Schichten beeinflussen die Feuchtkonvektion auf zweierlei Weise: Sie fungieren in erster Linie als Barrieren, die die Auslösung von Konvektion unterdrücken oder die Entwicklung zu hochreichender Konvektion verhindern. Ihre zeitliche Fortdauer kann zudem zu einer Anreicherung von Feuchte in der Grenzschicht führen. Haben sich die Lids aufgelöst oder üben sie ihren konvektionsunterdrückenden Einfluss nicht mehr aus, so kann sich, bedingt durch die akkumulierte Feuchte, hochreichende Konvektion ausbilden. Hochaufgelöste bodennahe Beobachtungsdaten ermöglichen eine genaue Lokalisierung der Konvergenzzonen, deren Rolle sich in Bezug auf die Auslösung von hochreichender Konvektion in allen untersuchten Fällen und selbst bei geringer atmosphärischer Unterdrückung als bedeutend erweist. Insgesamt konnte gezeigt werden, dass die Grenzschichtbedingungen die Auslösung von Konvektion stark beeinflussen: die Höhe der Konvektionsindizes wird durch den Grenzschichtzustand mit bestimmt und Konvektion wird in vielen Fällen durch Oberflächen- und Grenzschichtprozesse ausgelöst.

Es wird darüber hinaus ersichtlich, dass die Auslösung von hochreichender Konvektion stärker auf Konvergenz und eine niedrige CIN als auf eine besondere hohe CAPE angewiesen ist. Dagegen zeigt sich, dass für die Intensivierung konvektiver Systeme hohe CAPE-Werte und eine hohe Feuchte in der mittleren Troposphäre notwendig sind. Auf der Grundlage der Gesamtbewertung aller untersuchten CSIP-Intensivmessphasen wurden einige Schwellenwerte für das CSIP-Gebiet erstellt. Hochreichende Konvektion wird in Bereichen mit CAPE-Werten über 500 J kg^{-1}, absoluten CIN-Werten unter 50 J kg^{-1}, einer CAP-Stärke unter 2 °C (kennzeichnend für Lids in der Grenzschicht), einem Lid-Effect von ca. 0 °C (kennzeichnend für mitteltroposphärische Lids) und bei vorhandener bodennaher

Konvergenz beobachtet. Diese Schwellenwerte belegen, dass die konvektive Instabilität und die konvektionshemmenden Inversionen (CAP) während konvektiver Ereignisse in diesem Gebiet oft ziemlich schwach sind und dass die Auslösung über Grenzschichtkonvergenz erfolgt. Aufgrund der Unsicherheiten in den Berechnungen und der begrenzten Anzahl der untersuchten IOPs muss darauf hingewiesen werden, dass die Schwellenwerte eher als erster Hinweis denn als feste Richtwerte zu betrachten sind. Das abschliessend vorgestellte 2-D-Diagramm zur Vorhersage der Wahrscheinlichkeit von hochreichender Konvektion fasst die für das Auftreten von hochreichender Konvektion erforderlichen Bedingungen noch mal zusammen.

Abstract

The initiation of atmospheric convection is investigated using the synergy of different instruments. The impact of increased spatial data resolution on the detection of the initiation of deep convection is analyzed, and a methodology is developed and applied to determine the likelihood of deep convection over flat and complex terrains. With this purpose, Intensive Observation Periods (IOPs) are used from the Convective Storm Initiation Project (CSIP), which took place in southern United Kingdom (UK), and from the Convective and Orographically-Induced Precipitation Study (COPS), which was performed in southwestern Germany and eastern France.

Events with nonexistent or weak synoptic scale forcing were chosen to better understand the Planetary Boundary Layer (PBL) processes which play a critical role in the initiation of convective storms because it is along its discontinuities that convection often breaks out. To describe the likelihood of convection, convection-related indices are used representing atmospheric stability, convective available energy, inhibition and triggering mechanisms. The sensitivity of convection indices to the temperature and moisture variability observed in the PBL is investigated using high-resolution aircraft measurements. It is shown that using the maximum and minimum temperature and moisture observed along a flight track of approximately 36 km a difference of 680 J kg^{-1} in convective available potential energy (CAPE), and -150 J kg^{-1} in convective inhibition (CIN) is found. Thus, the representation of atmospheric conditions varied from low to high probability for deep convection. These measurements demonstrated that despite the dense radiosonde network deployed during the CSIP and COPS campaigns, the high spatial variability of the temperature and especially of the water vapour field in the boundary-layer cannot be resolved only with radionsonde data.

An increase of the spatial data resolution was obtained using the synergetic effect of data from the networks of radiosondes, Automatic Weather Stations (AWSs), synoptic stations (SYNOPs), and especially Global Positioning Systems (GPSs). Since the GPS measurements are retrievals of column-integrated water vapour mixing ratio, a new algorithm "adjustment method" is introduced to combine the IWV data from the GPS with the radiosonde profiles. This combined data set is used to calculate the spatio-temporal evolution of convection-related indices. The gained spatial resolution and the combination of convection-related indices presented successfully represented the areas where the likelihood of deep convection was high for several events with different dominant processes and for both, flat (CSIP) and complex terrains (COPS). Special emphasis is put on the investigation of the spatio-temporal evolution of low-level and mid-tropospheric stable layers or lids and their role in the inhibition of convection. The altitude, strength and persistence of these stable layers were observed to be important for the initiation of convection. Lids are seen to influence moist convection in two ways. Primarily, they act as a barrier that prevents the initiation of convection or constrains the development of deep convection. Furthermore, if lids persist in time, moisture can accumulate in the PBL, which eventually leads to deep convection when the lid is eroded or do not exert any constraining effect anymore.

High-resolution near-surface observations allowed a detailed localisation of convergence zones, whose role in deep convection triggering is seen to be decisive in all the investigated cases, even in cases where low atmospheric inhibition was observed. Location and timing of the initiation of convection were critically influenced by the structure of the humidity field in the boundary layer. Thus, it is demonstrated that PBL conditions exert a great impact on the initiation of convection because directly determine the convective indices, and surface and PBL processes often initiate convection. Moreover, it is seen that convergence and low CIN influenced the triggering of storms more than special high values of CAPE. However, the storm intensification was dominated by high CAPE and high mid-tropospheric moisture. From the joint evaluation of all CSIP IOPs investigated, some thresholds were established for the CSIP region. Deep convection is observed in areas where CAPE is higher than 500 J kg^{-1}, absolute values of CIN are below 50 J kg^{-1}, capping strength (CAP) is below 2°C (representing PBL-lids), lid effect is approximately 0°C (representing mid-tropospheric lids), and near-surface convergence exists. These thresholds confirm that convective instability and capping inversions in that region are often quite weak on convective occasions and triggering is controlled by boundary layer convergence lines. Nevertheless, these thresholds should be regarded as a reference because of the errors associated with the calculations and the limited number of IOPs investigated. Furthermore, a 2-D diagram representing the atmospheric conditions required for the initiation of convection is presented for the prediction of deep convection.

Contents

Chapter 1

Introduction

Deep, moist convection produces some of the most damaging weather phenomena. Heavy rain and consequent flooding is a direct consequence of convective storms. Damage caused by heavy rain is at the present time a major focus for international activities. Because of the severe damage caused by heavy precipitation and resulting flooding, there is a great need for reliable forecasting of storms. Storm events are often highly localized so that predicting why, where and when storms develop is one of the most difficult issues in weather forecasting. Predicting the location of the first outbreak accurately is crucial because once the first convective storm has formed, convection initiation preferentially occurs in the vicinity of that storm (Browning et al., 2007). The initiation and likelihood of deep moist convection is the general topic of this thesis.

Previous field campaigns and high-resolution modelling give us an idea of the processes which cause the triggering of deep convection (Carbone et al., 1990; Wilson et al., 1998; Wakimoto, 2001; Jorgensen and Weckwerth, 2003; Weckwerth and Parsons, 2006). Although processes involved are well understood in broad terms, the initial development of the convective cells remains inadequately understood (Bennett et al., 2006). One of the main goals of this thesis is to increase our knowledge about the initiation of precipitating convection allowing modellers to use such understanding in the development of improved Numerical Weather Prediction (NWP) models.

To the present day, operational numerical models often fail to predict the location and timing of convection initiation under weak or nonexistent synoptic forcing when triggering of convection results from boundary layer processes because these processes are not well represented in the models. However, the influence of surface and atmospheric boundary layer processes on the initiation of convection is often emphasised (Wilson et al., 1998) and boundary layer forcing has long been known to cause thunderstorm initiation (Purdom, 1982), so that detailed knowledge of the thermodynamics and dynamics of the boundary layer is required to predict the initiation of deep convection.

Former results suggested that convection initiation is very sensitive to variations in low-level temperature and moisture (Brooks et al., 1993; Crook, 1996). These spatial inhomogeneities are often bound to land use and/or soil water variability. These quantities control the energy balance of the earth's surface and result in inhomogeneities of the boundary layer temperature and moisture fields, which usually determine if and where the convective temperature is reached in the course of the day. Moreover, former studies demonstrated that near-surface or/and boundary layer inhomogeneities in the order of 10 km are necessary for generating mesoscale circulations that could trigger convection and release the instability (Schaedler, 1990). In particular, the low-level and boundary layer moisture field is a key factor to understand the initiation of precipitating convection (Lauscaux et al., 2004). Small changes in water vapour mixing ratio can control whether storms occur (Crook et al., 1991). In addition to advective processes, water vapour is made locally available in the atmosphere through evapotranspiration from soil and vegetation. Variations of about 1 g kg^{-1} in boundary layer moisture can make the difference between no initiation or intense convection (Crook, 1996). Weckwerth et al. (1996) showed that at least this magnitude of variability occurs routinely in the convective boundary layer (CBL) over a distance of only a few kilometers. That means that the spatial variability in the humidity field of the boundary layer air could be on scales too small to be resolved by conventional observations (Weckwerth, 2000). Moreover, convergence zones of different origin generated in the boundary layer often prove to be responsible for convection initiation (Fankhauser et al., 1985; Wilson et al., 1992, 1998; Kottmeier et al., 2008; Kalthoff et al., 2009). Thus, the understanding and successful prediction of storm initiation depends on an accurate specification of the initial PBL's thermodynamic and kinematic fields with particular attention to the water vapour field and triggering via PBL processes.

During the last years, two important meteorological experiments took place in Europe aiming to improve the knowledge about storm initiation: the Convective Storm Initiation Project (CSIP) (Browning et al., 2007), which was performed in the southern part of the United Kingdom in the summer 2005, and the Convective and Orographically-Induced Precipitation Study (COPS) (Wulfmeyer et al., 2008; Kottmeier et al., 2008), which took place in southwestern Germany and eastern France in the summer 2007. A key focus of these experiments was the observation of the CBL with a dense array of observing systems, including a radiosonde network, wind profilers, a network of global positioning systems (GPS) to retrieve water vapour content, and automatic weather stations (AWS) among others. An important difference between both experiments concerns the features of the landscapes where the experiments took place. Whereas the southern region of the UK is characterized by an intermediate level of orography and nearby coastlines whose marine environment largely influences the atmosphere, the COPS domain is characterized as a complex terrain where topography is a factor that largely influences the initiation of convection (Pielke and Segal, 1986; Banta, 1990).

During the CSIP campaign, from the 18 Intensive Operational Periods (IOPs) that took place during a three-month period, only one of the cases encountered was due to convection originating from mid and upper-tropospheric forcing; all of the other cases were due to convection initiating in the boundary layer (Browning et al., 2007). This makes this data set an exceptional opportunity to investigate episodes where the initiation of convection occurs via boundary layer processes. In this thesis, data from both measurement campaigns, CSIP and COPS, are used. This exceptional data base allows the investigation of the processes responsible for the initiation of convection in two completely different environments allowing the intercomparison of the findings in both types of terrains. Although the main focus is common to all the events investigated, in each case different processes dominate the atmospheric evolution.

In general, the typical resolution of conventional observing systems, such as radiosondes, is insufficient for observing the CBL variability. Several studies (Dabberdt and Schlatter, 1996; Weckwerth and Parsons, 2006) have suggested that a critical factor limiting the prediction of convective precipitation is the measurement uncertainty in the high-resolution distribution of water vapour. Accurate moisture measurements are critical, yet largely unavailable for many atmospheric science applications (Weckwerth et al., 1999). Radiosondes, the traditional means of obtaining water vapour measurements, are insufficient because they provide vertical profile information at widely distributed locations and are typically only available twice a day, sometimes containing significant errors and biases (Soden and Lanzante, 1996; Guichard et al., 2000; Wang et al., 2002; Turner et al., 2003). Even the dense network of radiosondes, especially deployed for CSIP and COPS, may be not dense enough to resolve near-surface and boundary layer variability that are expected to be relevant to the initiation of convection. Up to now, some technologies have been investigated to obtain a better representation of the water vapour field (Linne et al., 2007; Wulfmeyer, 1998). However, there is a general absence of operational, scanning ground-based remote sensing systems for water vapour so that this variable is not sampled, for example, on the same spatial and temporal scales as radar measurements of precipitation. Satellite techniques show great promise for covering these spatial and temporal scales; however, current systems cannot obtain high-vertical-resolution water vapour measurements over land with high accuracy in the lower troposphere and can often be limited by extensive cloud cover. GPS networks (Bevis et al., 1992; Ware et al., 1996; Wickert and Gendt, 2006) have a great potential for this application. The prospect of using GPS for measuring atmospheric water vapour for research and operational weather forecasting is promising because extensive networks of continuously operating GPS receivers exist around the world and the data could be made available at relatively little cost. In this thesis, a new algorithm is introduced to relate the GPS Integrated Water vapour (IWV) measurements with the boundary layer water vapour distribution obtained from radiosondes. By combining the radiosonde and GPS measurements, an increase in the resolution of the water vapour field is obtained. In addition, profiting from the high-resolution network of AWS available during CSIP, an increase in the resolution of the near-surface fields is obtained. Thus, one of the primary goals of this thesis is to investigate whether the radiosonde network is dense enough to trace back convection initiation to boundary layer conditions or if an increase in the resolution by AWS and/or GPS data leads to a significant improvement.

In addition to an accurate specification of the boundary layer thermodynamics, knowledge of the atmospheric conditions, such as stratification and the potential for inhibition, is required for the prediction of the likelihood of initiation of deep convection. Development of moist convection requires the existence of instability and a trigger mechanism like vertical motion. The type of convection, i.e. shallow or deep convection, depends on atmospheric stratification above the condensation level, the presence of capping inversions or lids, and the height of the equilibrium level. Lids are frequently observed features in most pre-storm environments (Carlson and Ludlam, 1968). These troposphere temperature inversions allow heat and moisture to accumulate in the boundary layer building-up an energy and humidity reservoir during the day for future thunderstorms. However, this inhibition must be overcome for a storm to form. The process of overcoming this inhibition normally controls if, where and when storms occur (Browning et al., 2007).

To characterize and quantify the above mentioned atmospheric conditions, and describe the likelihood of convection, convective indices are used (Lee and Passner, 1993; Fuelberg and Biggar, 1994; Huntrieser et al., 1997; Haklander and Van Delden, 2003; Kunz, 2007). Convective indices reflect the state of the atmosphere and help to quantify the potential for thunderstorm development according to the prevailing properties of the air mass. An increase in the resolution of the thermodynamic fields also results in an increase in resolution in the spatial representation of the convection indices and the convective inhibition parameters. In this thesis, convection indices are computed lifting a parcel with near-surface characteristics, surface-based (SB) parcel, or averaged over the lowest 50 hPa of the mixed layer (ML), 50 hPa ML parcel. This analysis allows the investigation of whether the parcels are - with respect to the observation of deep convection - better defined in terms of the near-surface properties or averaged over the lowest 50 hPa of the PBL. Many authors have investigated the efficiency of single convection indices derived from the observed vertical profiles for thunderstorm prediction (Schulz, 1989; Manzato, 2005; Kunz, 2007). In the last years, it has become clear that it is not enough to analyze individual convective indices to predict convective storms. As pointed out by former studies (Weckwerth, 2000; Graziano and Carlson, 1987), it is necessary to jointly evaluate several indices and parameters to ascertain the likelihood of deep convection. In this work, an attempt is made to find out an optimal combination of convection-related indices to describe the likelihood of deep convection based on the idea that a set of conditions is required for each weather phenomenon to occur (Doswell et al., 1996). The initial working hypothesis is that the likelihood for deep convection is high for regions with high conditional instability, low atmospheric inhibition, absence of mid-tropospheric lids, and the presence of a triggering mechanism. In this thesis this hypothesis is verified and a combination of convection related indices is defined to better represent the likelihood of deep convection. This also includes investigating the sensitivity of convection indices to the temperature and moisture variability in the CBL. Moreover, thresholds for the combined convection-related indices for the CSIP area are obtained.

This study shall contribute to further the understanding of how various measurements can be integrated to establish a method for the determination of the areas with higher deep moist convection likelihood. The general goals and the issues addressed in this thesis can be summarized as follows,

- Investigation of the parameters that control boundary layer variability and the scales in which these variations are dominant.

- Study of the effect that an increase in resolution at the near-surface and/or in the boundary layer has on a better detection of convection initiation.

- Derivation of a method to receive a resolution appropiate for detection of the initiation of convection by combining radiosonde and GPS data.

- Investigation of the most appropiate combination of convection related indices for prediction of deep convection likelihood.
- Definition of general thresholds for the convection-related indices used in the analysis of convection likelihood in the maritime environment of the UK.

This thesis is structured as follows: in chapter 2, a general introduction of atmospheric convection is presented. A description of the processes and the most significant parameters relevant for this study is given. In Chapter 3, the relation between the boundary layer processes and the initiation of convection is introduced. Chapter 4 presents the CSIP and COPS experiments. In Chapter 5, aircraft and radiosonde measurements are used to investigate the spatio-temporal variability of temperature and moisture in the boundary layer and the impact of spatial variability in the PBL on convection-related indices. The main purpose of Chapter 6 is to introduce the methods used and the algorithms developed to increase the resolution of the radiosonde-interpolated temperature and moisture fields using the synergy of different instruments. In Chapter 7, the spatio-temporal evolution of the obtained high-resolution fields of convection-related indices is investigated. A set of convection-related indices is presented to ascertain the likelihood of deep convection. In Chapter 8, the performance of the algorithms developed and the combination of convection-related indices are evaluated in four different IOPs from CSIP and COPS. In Chapter 9, all the results obtained for the different CSIP IOPs investigated are jointly evaluated in order to establish thresholds for the area of investigation. A new 2-D diagram representing the atmospheric conditions required for the initiation of convection is presented for the prediction of deep convection. Finally, the general conclusions are presented in Chapter 10.

Chapter 2

Atmospheric Convection

In general, convection refers to the transport of some property by fluid movement, most often with reference to heat transport. As such, it is one of the main processes by which heat and moisture is transported. In meteorology the term convection refers to transport by the vertical component of the flow associated with buoyancy. Progress in understanding convection has been made through field observations, laboratory experiments, analytical studies, and numerical modeling applied over a wide range of forms. In this chapter, the basic physics of convective storms is addressed.

2.1 Basic concepts of convection

Convection takes many forms in the atmosphere. Atmospheric convection can be either dry or moist, as well as being either shallow or deep. Dry convection means that there is no conversion of water vapour to liquid water as the parcel is displaced vertically, i.e., the parcel is sub-saturated. Moist convection occurs when there is conversion of water vapour to liquid water, i.e., the parcel becomes saturated. Moist convection, can release large amounts of latent heat, thus giving the storm the ability of self-organization and self-development. Shallow convection is defined as organized air motions with a depth of 1 to 2 km, the depth of the PBL.

Deep moist convection (DMC) compresses organized air motions which can extend up to approximately 10-15 km and produce a variety of hazardous weather events - large hail, damaging wind gusts, tornadoes, and heavy rainfall - which are generally the result of the energy released by phase changes of water.

Furthermore, convection can be classified as being free, resulting from lifting of the air mass due to thermal instability, or forced, resulting from lifting of the air mass due to some kind of dynamical mechanism. Lifting an air mass causes it to cool and forces saturation and condensation. Common forced mesoscale lifting processes are:

- frontal lifting: ascent at a narrow zone of transition between two air masses that differ in temperature and humidity. A warm front is the leading edge of an advancing warm and humid (less dense) air mass as it rides over a cold and dry (dense) air mass, while cold front is the leading edge of an advancing cold and dry air mass as it slides under a warm and humid air mass.
- orographic lifting: the forced rising of air up the slopes of a hill or mountain. As the air is forced to rise along the mountain's windward slopes (upwind side), it expands and cools, which increases its relative humidity. Meanwhile, on the mountain's leeward slopes (downwind side), air descends and warms, which reduces its relative humidity.
- wind convergence: occurs near the ground, generates upward motion as it lifts surface air to a certain level within the atmosphere.

The most easily seen, and hence most familiar form of convection in the atmosphere is the free, thermal convection that causes the cumulus cloud development. The horizontal length scale of these clouds is a few kilometers (Emanuel, 1994). Convective storms can take numerous forms. They range in scale from isolated thunderstorms involving a single convective cloud, or cell, to mesoscale convective complexes consisting of ensembles of multicell thunderstorms. In the next sections, a description of the processes and the most significant levels in the ascent of a parcel through the atmosphere is given.

2.2 Basic physics of convective storms

Atmospheric motions are governed by three fundamental physical principles: conservation of momentum, conservation of energy, and conservation of mass. Three basic laws are used to diagnose the physical processes in convective storms; they are Newton's equation of motion applied to a fluid (the "equations of motion"), the first law of thermodynamics applied to a fluid, and a statement of the continuity equation (Bluestein, 1993).

2.2.1 The equations of motion

The equations of motion are a set of hydrodynamical equations representing the application of Newton's second law of motion to a fluid system. The total acceleration on an individual fluid particle is equated to the sum of the specific forces acting on the particle within the fluid. Written for a unit mass of fluid in motion in a coordinate system fixed with respect to the earth, the vector equation of motion for the atmosphere is,

$$\frac{d\vec{v}}{dt} = -2\vec{\Omega} \times \vec{v} - \frac{1}{\rho}\vec{\nabla}p + \vec{g} + \vec{F}_r, \tag{2.1}$$

where $\bar{v}$=(u,v,w) is the three-dimensional velocity vector, $\vec{\Omega}$ is the rotation rate of the Earth about its axis, ρ is the density, p is the pressure, the centrifugal force has been combined with gravitation in the gravity term $\vec{g}$, and $\vec{F}_r$ designates the frictional force. This form of the momentum equation states that the acceleration following the relative motion in the rotating frame equals the sum of the Coriolis force, the pressure gradient force, the effective gravity, and friction. The equation of motion can be separated into the horizontal and the vertical components. In both, the equations of horizontal and vertical motion, the molecular viscosity terms may be neglected when considering air motion not right next to the surface.

The vertical momentum equation for an ideal fluid is written as,

$$\frac{dw}{dt} = -\frac{1}{\rho}\frac{\partial p}{\partial z} - g. \tag{2.2}$$

The density and pressure can be divided into a standard density/pressure, p_0(z) and ρ_0(z), which are the horizontally averaged density/pressure at each height and the deviations from the standard values of pressure and density, p′ and ρ', which would be zero for an atmosphere at rest. The horizontally averaged density/pressure are defined so that p_0(z) and ρ_0(z) are in *exact* hydrostatic balance, $1/\rho_0(dp_0/dz) \equiv -g$, i.e. the downward force on a parcel due to gravity is exactly balanced by the upward force due to the pressure gradient, All vertical accelerations arise from the perturbations. Thus,

$$p(x,y,z,t) = p_0(z) + p'(x,y,z,t), \tag{2.3}$$

$$\rho(x,y,z,t) = \rho_0(z) + \rho'(x,y,z,t). \tag{2.4}$$

Expanding the inverse density and inverse pressure in a Taylor series and dropping all terms of second and higher order, the equation of vertical motion can be written as follows,

$$\frac{dw}{dt} = -\frac{1}{\rho_0}\frac{\partial p'}{\partial z} - g\left(\frac{\rho'}{\rho_0}\right), \tag{2.5}$$

where w is the vertical component of the wind. The first term on the right is usually referred to as the nonhydrostatic pressure gradient acceleration, which generally arises from dynamical effects of forced momentum changes, and the second term on the right is the buoyancy acceleration which represents the action of gravity on density anomalies,

$$B \equiv -g \left(\frac{\rho'}{\rho_0} \right). \tag{2.6}$$

For moist, unsaturated air the equation of state can be written as, p=$\rho R_d T_v$, where R_d is the gas constant for dry air and T_v is the virtual temperature. T_v is defined as, T_v=T(1+(R_v/R_d - 1)r)$\approx$T(1+0.608r), where R_v is the gas constant of water vapour and r is the water vapour mixing ratio. Using Eq.2.3 and Eq.2.4 for the pressure and density, respectively, and in the same way T_v(x,y,z,t)= $T_{v,0}$(z)+T'_v(x,y,z,t), one may obtain the relation ρ'/ρ_0=p'/p_0 - $T'_v/T_{v,0}$. The contribution of pressure variations to buoyancy may usually be neglected for flows in which the maximum velocity variations are substantially subsonic (Emanuel, 1994). Therefore, substituting in Eq.2.6 leads to,

$$B \simeq g \left(\frac{T'_v}{T_{v,0}} \right). \tag{2.7}$$

For large-scale motions, i.e. those for which the horizontal scale is much longer than the vertical scale, which is in the order of the depth of the troposphere ($\approx$10km), there is no significant buoyancy force, vertical air parcel accelerations are negligible, and the vertical pressure-gradient force is exactly counteracted by gravity (Ogura and Phillips, 1962; Emanuel, 1994).

2.2.2 Thermodynamics

The first law of thermodynamics applied to the atmosphere is an expression of the more universal physical law of the conservation of energy. From it, fundamental quantities for the study of the atmosphere can be derived.

The potential temperature, θ, is the temperature that a parcel of dry air at pressure p and temperature T would have if it were expanded or compressed adiabatically (i.e., a reversible process in which no heat is exchanged with the surroundings) to a standard pressure, p_{00}, usually 1000 hPa. Thus, every parcel has a unique value of θ, and this value is conserved in adiabatic displacements of unsaturated air.

$$\theta \equiv T \left(\frac{p_{00}}{p} \right)^{R'/c'_p}, \tag{2.8}$$

where R$'$ is the effective gas constant, R$'$=c'_p-c'_v. c'_v=c_{vd}(1+0.94r) and c'_p=c_{pd}(1+0.85r) are the specific heats of the air sample at constant volume and constant pressure, respectively. The quantities c_{vd} and c_{pd} are the corresponding values for dry air. To take into account the effect of water vapour dependence of density, one defines the virtual potential temperature, θ_v, given by Emanuel (1994),

$$\theta_v = T_v \left(\frac{p_{00}}{p}\right)^{R_d/c_{pd}}. \tag{2.9}$$

Since r and θ are conserved in adiabatic reversible processes, θ_v is a conserved variable in unsaturated air and is also directly related to density (Emanuel, 1994).

When a parcel of unsaturated air is lifted adiabatically, it expands and cools, conserving its virtual potential temperature. The rate at which its temperature decreases with height is called the dry adiabatic lapse rate, Γ_d. From the adiabatic form of the first law of thermodynamics and the hydrostatic equation, the dry adiabatic lapse rate taking into account the small effect of water vapour is defined as,

$$\Gamma_d = -\left(\frac{dT}{dz}\right) = \frac{g}{c_{pd}}\frac{1+r}{1+r\left(\frac{c_{pv}}{c_{pd}}\right)} \approx 10Kkm^{-1}, \tag{2.10}$$

where r is conserved during the ascent. The dry adiabatic lapse rate is much larger than the typically observed environmental lapse rate, Γ_e, in the troposphere of $\approx$ 6-7 K km^{-1}. The difference is mostly due to the presence of water and its phase changes.

If a parcel of moist air is forced to rise, it will eventually become saturated. The level at which saturation occurs is called the lifting condensation level (LCL) (Fig.2.1). Following Bolton (1980), the temperature of the LCL, T_{LCL}, is defined as,

$$T_{LCL} = \frac{2840}{3.5\ln T - \ln e - 4.805} + 55, \tag{2.11}$$

where T is in K and water vapour pressure e in hPa.

When parcels are lifted by buoyant forces, associated with surface heating, saturation occurs at the convective condensation level (CCL). The CCL corresponds to the LCL for a layer which is already well-mixed. Discussion of parcel dynamics in a moist atmosphere is facilitated by defining a thermodynamic variable called the the pseudo-equivalent potential temperature, θ_{ep} (Bolton, 1980),

$$\theta_{ep} = T\left(\frac{p_{00}}{p}\right)^{0.2854(1-0.28r)} exp\left(r(1+0.81r)\left(\frac{3376}{T_{LCL}} - 2.54\right)\right). \tag{2.12}$$

The moist adiabatic lapse rate in this case called pseudo-adiabatic lapse rate is,

$$\Gamma_s \equiv -\frac{dT}{dz} = \Gamma_d \frac{1+\frac{L_v r}{R_d T}}{1+\frac{L_v^2 r(1+r/\varepsilon)}{R_v T^2(c_{pd}+rc_{pv})}}, \tag{2.13}$$

where $\varepsilon = R_d/R_v$. Γ_s is always less than Γ_d. Observed values of Γ_s range from $\approx$ 4 K km^{-1} in warm humid air masses in the lower troposphere to $\approx$ 6-7 K km^{-1} in the midtroposphere (Holton, 1992). Therefore, if the parcel would be forced to rise pseudo-adiabatically from the LCL, its state would follow along the pseudo-adiabat passing through the LCL. A pseudo-adiabatic process can be defined as an irreversible process in which the heat capacity of liquid water (or ice) is neglected, i.e. all condensation products are assumed to fall out. However, one can also assume that the condensed water remains inside the parcel. That assumption results in the moist adiabatic lapse-rate, Γ_m. In this reversible moist adiabatic transformations, the moist entropy represented by the equivalent potential temperature, θ_e, is conserved. The difference between an irreversible pseudo-adiabatic and a reversible moist adiabatic ascent is small in the lower and mid-troposphere. However, θ_e is a function of pressure, temperature, and liquid water content, thus, it cannot be represented in a single two-dimensional thermodynamic diagram. In this thesis all related calculations assume pseudo-adiabatic ascent of moist saturated parcels. Sometimes isopleths of θ_{ep} are labeled in terms of their temperature at the reference pressure of 1000 hPa. The quantity is then referred to as wet bulb potential temperature, θ_w. If the environmental lapse rate is greater than the saturated adiabatic lapse rate, the parcel becomes warmer than its environment and buoyant relative to its surroundings, just above where the pseudo-adiabat crosses the temperature sounding. This level is called the level of free convection (LFC) (Fig.2.1). Thereafter, the parcel cand rise freely under its own positive buoyancy until it reaches the equilibrium level (EL) (Fig.2.1). Above this level, it may become negatively buoyant and decelerates. The EL is close to the height of the top of a cumulus cloud, if subcloud layer air were first able to reach its LFC.

2.2.3 Conservation of mass

The equation of continuity, which is a statement for the conservation of mass for a compressible atmosphere is defined by,

$$\left(\frac{1}{\rho}\right)\frac{d\rho}{dt} + \vec{\nabla}\cdot\vec{\mathrm{v}} = 0. \tag{2.14}$$

To a good approximation the compressibility of the atmosphere may be ignored in the equation of continuity if the air motions in the atmosphere are relatively shallow. This approximation is possible because the basic state (standard atmosphere) density varies across the lowest kilometer of the atmosphere by only about 10%, and the fluctuating component of density deviates from the basic state only a few percent (Holton, 1992). Thus, the so-called Boussinesq approximation can be used to describe air motions for shallow cumulus clouds and the boundary layer, the region near the ground that is affected by it, typically up to ≈ 1 km above ground level (AGL), but up to as much as 3 km AGL or more in heated, arid regions. *"The Boussinesq approximation neglects density variations in the fluid except when they are coupled with gravity"* (Emanuel, 1994). Therefore, Eq.2.14 reduces to,

$$\vec{\nabla} \cdot \vec{v} = \frac{\partial u}{\partial x} + \frac{\partial v}{\partial y} + \frac{\partial w}{\partial z} = 0. \tag{2.15}$$

Eq.2.15 provides one mean of obtaining the wind vertical component indirectly. The horizontal components of wind can be measured directly. The vertical component, despite its great potential interest to meteorologists, is far too small to be measured accurately by the usual meteorological measuring systems. It has to be inferred indirectly. Thus, re-arranging the incompressible continuity equation into the form,

$$\frac{\partial w}{\partial z} = -\left(\frac{\partial u}{\partial x} + \frac{\partial v}{\partial y}\right), \tag{2.16}$$

the term on the right is called Şhorizontal divergenceŤ. If we know the horizontal divergence at several levels in the atmosphere, then the vertical velocity at any level can be calculated by integrating the continuity equation,

$$w = -\int_0^z \left(\frac{\partial u}{\partial x} + \frac{\partial v}{\partial y}\right) dz. \tag{2.17}$$

This result assumes that w=0 at z=0. It turns out that this result is not as useful as it might appear. This is due to the fact that the horizontal divergence can rarely be estimated to a good degree of accuracy. It is nearly always the small difference between two relatively large, nearly equal quantities, $\partial u/\partial x$ and $\partial v/\partial y$. The errors in estimating the horizontal components of the wind, usually $\pm$1-2 m s^{-1}, mean that the errors in the horizontal divergence often approach 100%, and so the vertical velocity can only be estimated very crudely.

2.3 Atmospheric stability

One of the requirements for the initiation of atmospheric deep convection is the existence of some kind of instability. The atmospheric stability is the property of the environmental (ambient) air that either suppresses (stable) or enhances (unstable) the vertical motion of an air parcel. Therefore, to assess whether any kind of instability is present, the atmospheric stability must be investigated.

Atmospheric stability can be characterized conveniently regarding the oscillation frequency of an infinitesimal parcel set in motion by a small perturbation, δz

(Emanuel, 1994). This is called the Brunt-Väisälä, or buoyancy, frequency. The higher the oscillation frequency, the more stable the air to vertical displacements. The buoyancy frequency becomes imaginary as the air becomes unstable. Assuming in Eq.2.5 that the perturbation pressure gradient accelaration is negligible for the small dimensions of the displaced parcel, i.e. p'=0, the vertical acceleration of the parcel can be written as,

$$\frac{dw}{dt} = B. \tag{2.18}$$

For an infinitesimal displacement, δz, Eq.2.18 becomes,

$$\frac{d^2\delta z}{dt^2} - \frac{\delta B}{\delta z}\delta z = 0, \tag{2.19}$$

where $\delta B/\delta z$ is the change of the parcel's buoyancy with height. Eq.2.19 is an oscillator equation, giving the oscillation frequency as,

$$N^2 = -\frac{\delta B}{\delta z}, \tag{2.20}$$

where N is the buoyancy or Brunt-Väisälä frequency.
Re-writing Eq.2.7 in terms of θ_v for moist unsaturated air (for dry unsaturated air it would be written in terms of θ),

$$B = g\left(\frac{\theta_v'}{\theta_{v,0}}\right), \tag{2.21}$$

where $\theta_{v,0}$ and θ_v' are the standard and the deviation from the standard virtual potential temperature, respectively. Then, the buoyancy frequency can be written as,

$$N^2 = \frac{g}{\theta_{v,0}}\frac{\partial\theta_{v,0}}{\partial z}. \tag{2.22}$$

This expression shows that the convective stability of moist unsaturated air is proportional to the vertical gradient of θ_v (for dry unsaturated air, it is proportional to the vertical gradient of θ). An analogous expression (not shown) (Emanuel, 1994) can be obtained for the buoyancy frequency of cloudy or saturated air taking into account that vertical displacements of saturated air will result in phase change of water, which affects the buoyancy of the air. This expression shows that saturated air is unstable when θ_e decreases upward and/or total water, r_T, increases upward.

While the stability classification obtained from this approach is useful for assessing stability within clouds, a different technique must be used for evaluating the stability of unsaturated atmospheres to the development of clouds themselves. A general method of assessing the stability of the atmosphere to convection is to evaluate the buoyancy of a parcel displaced a finite distance under a reversible or pseudo-adiabatic process, this is the so-called parcel method. The parcel method presents some disadvantages as for example that it can not account for the reaction of the environment to the parcel displacement, i.e. it does not account for perturbation pressure gradient accelerations. Therefore, although the parcel method is not without its problems, we are faced with using it to assess instability, until a new method is developed (Schultz et al., 2000). Nevertheless, the use of this method brings significant advantages as the aforementioned ability to ascertain the ability of an unsaturated parcel to become buoyant, the possibility of measuring the potential energy for convection, as well as the atmospheric inhibition present, both later discussed, and the ability to assess the existence of conditional instability.

For any one parcel that is unsaturated, two points are needed to describe its state on a thermodynamic diagram. One describes the temperature and the other describes the humidity. A Skew T-Log p (Fig.2.1) is a "pseudo-adiabatic thermodynamic diagram" in which the "parcel method" is used to determine unmeasured parameters and to analyze the characteristics of a vertical profile of the troposphere and lower stratosphere above a station. Conceptually, parcel theory describes the vertical motion and changes of state of the well-defined parcel of air in the atmosphere. According to the simple parcel theory, convection begins in a layer, close to or over the surface, where an air parcel is lifted by either buoyant or mechanical forces, causing it to cool off as its pressure decreases and it expands. Eventually, it cools sufficiently so that the water begins to condense into droplets, forming a cloud. This releases latent heat, which warms the parcel; under certain conditions, this can increase the parcel's buoyancy sufficiently so that the parcel continues to rise. Eventually, the condensate grows to the size that it can precipitate out.

When plotted with the various meteorological elements received from an upper air sounding, the Skew T-Log p diagram presents a vertical picture of the atmospheric conditions present at the time of observation and allows for computations of significant levels and various parameters required by forecasters. Two characteristic features of this thermodynamic diagram are that the horizontal isobars are spaced at equals increments of log(p), so that the vertical ordinate is approximately proportional to height, and isotherms have been skewed to the right. This diagram represents pressure, density, temperature, and moisture in such a way that the basic atmospheric energy transformations are visually depicted. A unit of area on the diagram represents a specific quantity of energy. So an important advantage of this method is the capability of measuring the total reservoir of potential energy for the convection of a parcel (CAPE) and the energy that needs to be overcome by an air parcel to reach its LFC (CIN).

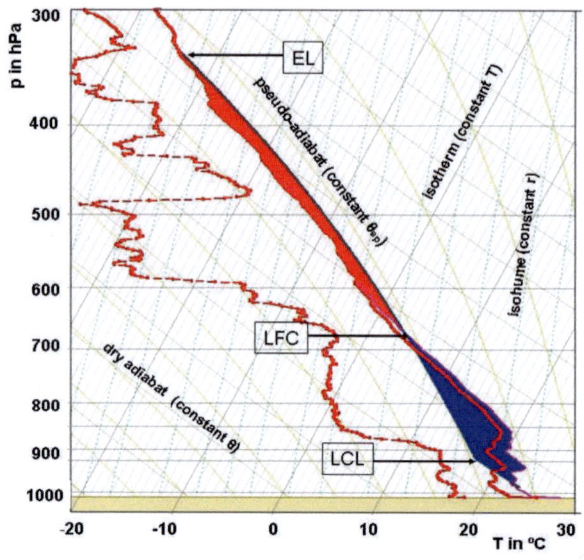

Figure 2.1: Skew T-log p diagram. The red profile on the left indicates the dew-point temperature while that on the right indicates the temperature. The pink profile on the right of the temperature profile indicates the virtual temperature. Some of the most significant levels in the atmosphere are indicated as: LCL (Lifted Condensation Level), LFC (Level of Free Convection), and EL (Equilibrium Level). The area highlighted in red, between the LFC and the EL, corresponds to the Convective Available Potential Energy (CAPE) and the area in blue, between the level of the lifted parcel and the LFC, is the Convective Inhibition (CIN). In the background of the diagram, the solid diagonal lines slanting upward and toward the right are isotherms, the dashed lines slanting less sharply toward the right than the isotherms are the saturation mixing ratio lines, the solid lines slanting upward and to the left are dry adiabats, and the curved solid lines are pseudo-adiabats.

As a summary, according to the concept of atmospheric instability discussed in this section, a general classification is as follows (Tab.2.1).

- *Absolute instability*, is a property of an ambient air layer that is unstable for both saturated and unsaturated air parcels. It occurs when the temperature of the ambient air deacreases more rapidly with altitude than the dry adiabatic lapse rate. In other words, a parcel rising from the surface, cooling at either

	definition	additional requirement
ABSOLUTE INSTABILITY	$\partial T/\partial z < \Gamma_d$ or $\partial\theta/\partial z < 0$	
POTENTIAL INSTABILITY	$\partial\theta_e/\partial z < 0$	the layer must be lifted to saturation
CONDITIONAL INSTABILITY	$\Gamma_d < \partial T/\partial z < \Gamma_s$ or $\partial\theta_{es}/\partial z < 0$	the release of conditional instability requires CAPE>0

Table 2.1: Atmospheric instability classification. θ_{es} is the saturation equivalent potential temperature.

the dry adiabatic or moist adiabatic rate will be warmer than the surrounding environment, and will continue to rise on its own.

- *Potential instability*, the state of upward-decreasing θ_e is one of potential instability, also called convective instability. The instability is potential in the following sense: "Were the entire air mass lifted bodily until it becomes saturated, then the upward decrease of θ_e does imply instability ". Therefore, at some point during the lifting process, the air mass will become unstable in the parcel sense (Emanuel, 1994).

- *Conditional instability*, means that the environment is locally stable to the ascent of unsaturated air parcels, but unstable to the ascent of saturated air parcels. Conditional instability is a special case of finite-amplitude instability, a general condition in which a system is stable to small perturbations but unstable to at least one type of perturbation of sufficient amplitude. The release of conditional instability requires not only that $\partial\theta_{es}/\partial z < 0$ for a saturated parcel, but also the presence of an air parcel that must reach the LFC. A major difference between potential and conditional instability is that lifting does not affect whether a layer is potentially stable or unstable with respect to vertical displacements of the layer, but lifting can destabilize a column of the atmosphere with respect to vertical parcel displacements in such a way that a stable column becomes unstable. The two types of moist instability are not mutually exclusive: before saturation occurs a sounding can exhibit both types. In fact, soundings with conditional instability generally have $\partial\theta_e/\partial z < 0$ at least in the lowest kilometer or so; therefore, they also have potential instability (Banta, 1990).

The terminology used for establishing a classification scheme for stability can be very misleading today. Definitions of conditional instability commonly used include both the lapse-rate definition (i.e., the environmental lapse rate lies between the dry- and the moist-adiabatic lapse rate) and the available-energy definition (i.e., a parcel possesses positive buoyant energy, CAPE>0), neither of which can be considered an instability in the classic sense. The stability of a layer (as defined by its lapse rate) can differ dramatically from the parcel stability (as defined by its buoyancy). Some authors make the distinction between these two possible definitions using the term "conditional instability" for the lapse-rate definition and the term "latent instability" for the available-energy definition (Normand, 1931a,b; Schultz et al., 2000). Clearly, if a sounding has no layer of conditional instability (lapse-rate definition), deep, moist convection is precluded (i.e., there can be no CAPE if lapse rates do not exceed moist adiabatic somewhere in the environmental sounding). In contrast, the existence of an atmospheric layer with lapse rate between the dry- and the moist adiabatic lapse rate does not ensure that CAPE>0. For the purpose of this work, "conditional instability" will hereafter refer to the available-energy definition.

2.4 The role of lids

As defined by Browning et al. (2007), a lid is a layer of warm, dry air that traps air of high θ_w, usually in the boundary layer, beneath potentially colder air in the middle and upper troposphere. The lid has been recognized for some time as a frequently observed feature of the pre-severe storm environment. It was introduced by Carlson and Ludlam (1968) by a conceptual model that described how the combined effects of topography, surface heating, and large-scale weather pattern can produce a lower-tropospheric inversion that leads to an enhancement of conditional instability and, therefore, to deep cumulus convection. They called this inversion a ŞlidŤ because of its effect in suppressing the release of the conditional instability. Paradoxically, the severity of the thunderstorms is increased by the presence of a lid because a large conditional instability can increase with time beneath the lid. Convection is more severe where a strong lid was present, although it is nevertheless more likely to occur where the lid is absent (Graziano and Carlson, 1987).

The so-called "lid sounding" (Fig.2.2) is characterized by a moist layer at the lowest levels with high values of wet bulb potential temperature, above which lies an abrupt relative humidity break and coincident thermal inversion (Fawbush and Miller, 1952). Since the inversion is usually found just above a level marked by a sharp decrease in relative humidity with altitude, relative humidity break , this

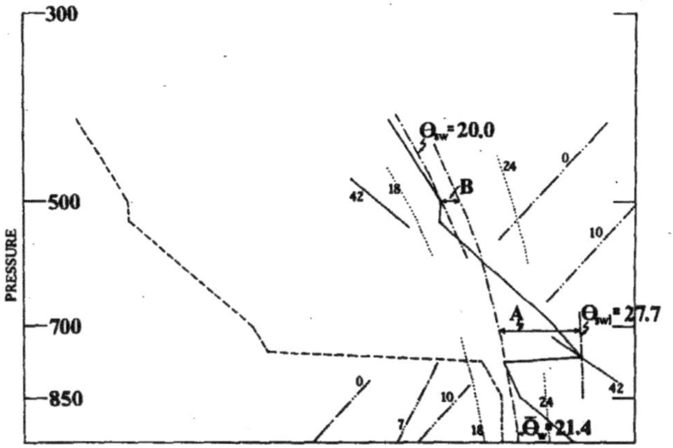

Figure 2.2: Schematic lid sounding on a Skew T-log p diagram. The vertical temperature profile is indicated by the solid line and that of the dewpoint by the dashed line. The dash double-dotted line segments slanting upward and toward the right are isotherms (labelled in °C). The dash triple-dotted line segment slanting less sharply toward the right than the isotherms is a portion of a saturation mixing ratio line. The solid line slanting upward and to the left is a dry adiabat. The curved dotted lines are moist adiabats. The dash single-dotted lines are moist adiabats representing the mean wet-bulb potential temperature of a layer between 30 and 80 hPa above the ground ($\bar{\theta}_w$) and the saturation wet-bulb potential temperature (θ_{sw1}) at the base of the elevated mixed layer. The saturation wet bulb potential temperature (θ_{sw}) of the air at 500 hPa is also indicated. The letters A and B refer to temperature differences corresponding, respectively, to the lid effect and buoyancy terms in Eq.2.31. (Adapted from Graziano and Carlson (1987)).

feature enables us to identify objectively the presence of a lid since the decrease with height of relative humidity near the lid is typically in excess of more than 1% per millibar. A unique aspect of the lid concept is that it accounts for the presence of inversions between 850 and 500 hPa. In this work, the criteria introduced by Graziano and Carlson (1987) are used to identify the presence of a lid and the associated relative humidity break in a sounding. The criteria for identifying the relative humidity break are:

- The relative humidity decreases with height by at least 1% hPa^{-1} between two successive significant levels not farther than 100 hPa apart.

- The lower of the two significant levels is considered the level of the relative humidity discontinuity, and a search is made above that level for the lid.
- The base of the dry layer is below 500 hPa.

The criteria for identifying the lid are:

- The sounding shows an increase in temperature or an isothermal layer within 100 hPa above the level of the relative humidity break; this criterion ensures the existence of large vertical gradients of moisture and potential temperature, which mark the boundary between the conditionally unstable air with high values of mean wet bulb potential temperature and the elevated mixed layer above.
- The lid base is the first significant level at the top of the isothermal or inversion layer.
- The lid is below 500 hPa.

Lid formation is generally due to differential advection of warm, dry air from an arid (and usually elevated) region over moist terrain, where the mixing layer is likely to be relatively shallow and the values of wet bulb potential temperature near the surface relatively large (Graziano and Carlson, 1987). Dealing with the initiation of convection, it is important to be able to observe the detailed structure and evolution of the lids because this process of overcoming the inhibition normally controls if, where and when storms occur. The area of significant convection on any given day is relatively small in comparison with regions where there is conditional instability. Severe thunderstorms are unlikely to occur, even in regions where the conditional instability is large, if a strong lid is present. A lid assists in the build-up of conditional instability by allowing warm, moist air to accumulate at low levels, thereby increasing the available potential energy. To initiate deep convection, the low-level air must eventually penetrate the lid. One way to achieve this is through progressive warming and/or moistening of the low-level air, perhaps as part of a diurnal trend. During the day, the boundary layer generally warms, becoming more buoyant or unstable. If sufficient warming occurs, the temperature inversion weakens, perhaps disappearing entirely. Boundary layer parcels may be then sufficiently buoyant to rise to the LFC, resulting in convection and possibly thunderstorms. Therefore, the location of the initial outbreak will be influenced partly by any spatial variability in temperature and/or humidity of the boundary layer air. Sometimes the variability is due to the effects of the variable terrain height or differing land (or sea) surface characteristics, or perhaps differential shadowing by clouds at higher levels. Another important factor is the variability in the strength of the lid itself. As described by Browning et al. (2007), this can be either intrinsic variability owing to the differing source regions for different parts of the lid, or local variability owing to some mesoscale dynamical mechanism that lifts the lid locally. Such lifting will cool the lid rapidly (at the dry adiabatic rate), thereby eliminating the inhibition and enabling boundary layer air of high θ_w (ascending moist adiabatically) to penetrate it (Fig.2.3). A small vertical displacement of the lid may be sufficient to allow the underlying air to penetrate buoyantly upward. Local lifting often occurs associated with convergence lines. In the morning hours, a challenge in severe weather forecasting is to determine not only if, but also where the lid may break later in the day, assuming the model represents adequately the lid. Model forecasts help determine future soundings. Based on the model data, severe storm indices can be calculated that help to measure the predicted dynamic forcing, instability, and the future state of the lid. It is important to remark that calculations of the lid are meaningless unless there is instability. If there is no instability, there will be no convection.

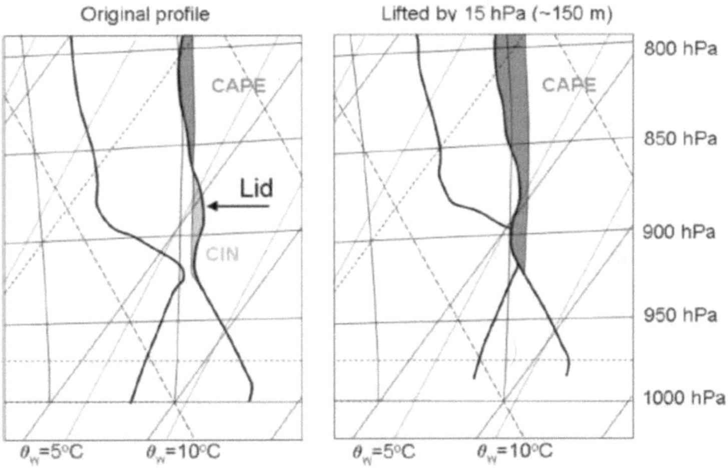

Figure 2.3: Illustration of how adiabatic lifting of a profile by as little as 15 hPa (≈150 m) can increase the available potential energy (indicated by CAPE) and completely eliminate the inhibition (indicated by CIN). The original profile is shown on the left while the lifted one is displayed on the right. (Adapted from (Morcrette et al., 2006).)

2.5 Convection-related indices

The so-called ingredients-based methodology states that three conditions are required for deep, moist convection: lift, instability and moisture (Fig.2.4). Sherwood (2000) argues that the condition of instability is unfortunately labeled since the three ingredients are actually attempting to diagnose the instability, of which the visible manifestation is deep, moist convection. The use of the term instability among the conditions discussed above was intended to infer the lapse-rate definition of conditional instability (Doswell III, 1987). The available-energy definition of conditional instability, sometimes called latent instability, combines two of the main conditions necessary, instability and moisture (Normand, 1931a,b; Schultz et al., 2000). The three conditions were chosen i) to imply the presence of convective available potential energy (CAPE) via conditional instability and moisture, and ii) to realize that convective potential via the ascent of parcels to their LFCs. The degree of stability affects how severe the convection can be, and an indication of the lapse rate between the lower and middle troposphere is an ingredient (explicitly or implicitly) of many of the severe weather indices.

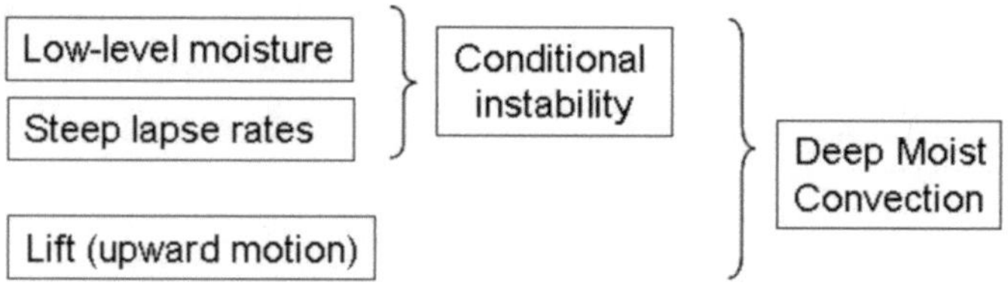

Figure 2.4: Required conditions for the initiation of deep moist convection (John and Doswell III, 1992).

In addition to the aforementioned conditions, it is also important to take into account the atmospheric inhibition. Very high atmospheric inhibition will not allow thunderstorms to break out (parcels will not be able to reach their LFC), and very low inhibition may lead to the continuous release of latent and sensible heat in small clouds, preventing sufficient energy from building up to drive strong thunderstorms. For quantifying the atmospheric conditions convection indices are used. The convection indices are thermodynamic and kinematic parameters which reflect the potential for thunderstorm development according to the prevailing properties of the air mass. The importance of computing stability indices in the course of analyzing sounding data to ascertain the atmospheric conditions is well known (Doswell, 1982; Miller, 1975). In many studies, the efficiency of various indices derived from the observed vertical profiles for thunderstorm prediction was investigated, for example by Schulz (1989) or more recently by Manzato (2005). Also, some studies have been made to evaluate different convection indices with respect to their skills and efficiency to predict thunderstorms of variable severity (Kunz, 2007).

Convection indices take into consideration various meteorological parameters, such as instability, moisture, shear ... No single index works best in all locations, or for all types of severe weather (wind, large hail, heavy precipitation or tornados) (Doswell III, 1987). For the calculation of the convection indices there is a multitude of options one can chose from to compute them. For example, the parcel choice, the use or not of the virtual temperature correction (Doswell III and Rasmussen, 1994) or the choice between several measures of a meteorological parameter (e.g., instability or vertical wind shear) which are calculated using different convection indices whose computation can lead to different interpretations. For example, the conditional instability can be quantified by the Convective Available Potential Energy (CAPE) of a parcel, which is a vertical integrated measure, or alternatively, by indices that represent the thermal buoyancy of a parcel at some altitude, such as the Lifted Index (LI) (Galway, 1958). The parcel choice implies that in the computation of the convection indices one may chose to lift the surface-based (SB) parcel, the most unstable (MU) parcel in the lower atmosphere, or a mixed-layer parcel (ML) of some predetermined depth. Even in this last case one may chose the depth of the parcel one wants to lift, 10hPa mean, 50hPa mean, 100 hPa mean are the most common choices. The SB parcel choice uses the surface air and dewpoint temperatures to determine the parcel ascent path. Although it allows a better representation of surface-based convection, calculations can be highly dependent on small time and space scales when these thermodynamic parameters display significant variations. The MU will always produce the largest estimate of buoyancy among the three measures introduced above. It is most effectively used to asses the potential for thunderstorms that develop in an unstable air mass situated above a stable surface layer, the so-called elevated convection (Rochette et al., 1999). In such cases, the SB and ML calculations will be relatively small (perhaps zero) in the computation of CAPE, thus failing to reveal the true convective potential. The ML parcel choice is used lifting a parcel constituting a well-mixed layer of constant potential temperature and mixing ratio. This layer is usually surface-based, but it may also be an elevated layer, this being a fourth variant of the parcel choice. There is not a universally accepted depth of the surface-based layer to mix. For example, Bluestein and Jain (1985) used the lowest 500 m layer, weighted by pressure, to compute the MLCAPE; Wakimoto and Wilson (1989) used the lowest 50 hPa layer; Johns et al.(1993) used the lowest 100 hPa layer; Rasmussen and Blanchard (1998) and Weisman and Klemp (1982, 1984) used the lowest 1000 m layer. The 100hPa/50hPa and 1000m/500m layers are quite similar in depth, since to a first-order approximation, 1 hPa $\approx$ 10 m near the earth's surface. Due to the averaging properties of the ML parcel choice, it is less variable in time and space than SB. Values of ML are typically smaller than those of SB. The ML and SB will be equal when the boundary layer is well mixed throughout the lifting layer. There is some debate whether SB or ML gives a better estimate of surface-based deep moist convection. In cases of shallow moisture (or perhaps erroneous surface data), the SBCAPE may be much larger than the MLCAPE, perhaps leading to an overestimation of the buoyancy. When a shallow surface-based inversion is present, the SBCAPE will be considerably lower than the MLCAPE, this time potentially underestimating the buoyancy. In these cases, it appears that MLCAPE is superior to the SBCAPE. A climatology of cloud-base heights in comparison to sounding-derived LCL heights would be useful. It is also very important to note that the virtual temperature correction can produce significant differences in the computation of some convection indices and derived parameters (Doswell III and Rasmussen, 1994).

The SB and ML parcel methods are chosen for the analysis. The SB parcel is chosen to estimate the influence of near-surface conditions in the initiation of convection, and the ML parcel is chosen to estimate the effect of using a representative value of temperature and moisture from the well mixed PBL in the prediction of convection initiation. Among the various possibilities concerning the parcel depth in the ML parcel choice, the 50 hPa (500 m) layer is used since this has been the most common choice in former studies, and the well-known thresholds for the convection indices are based on calculations made using this choice. Moreover, the virtual temperature correction is applied.

The convection-related indices used in this study can be divided into three representative groups in agreement with the identified conditions necessary for the initiation of convection: a) Available energy for convection and atmospheric stability, b) convective inhibition, and c) triggering. In the first group, the moisture at low levels and the atmospheric stability are quantified using CAPE and LI, which quantify the atmospheric **conditional instability**. The CAPE is an integrated value, whereas the LI is calculated at a characteristic atmospheric level. Therefore, the purpose of investigating the two different indices despite the fact that both quantify the conditional atmospheric instability lies in comparing the performance and differences resulting from the different ways of computation. In addition, the KO-index is used to quantify the atmospheric **potential instability**. Moreover, the Bulk

Richardson Number (BRN) is used to take into account **shear**. In the second group, the **atmospheric inhibition** is quantified making use of the well-known Convective Inhibition (CIN), the Capping inversion strength (CAP), and the Lid effect index. The last two quantify the strength of low- and mid-tropospheric inversions, respectively. In addition, the Lid Strength Index (LSI) which quantifies the convective potential of the atmosphere when mid-tropospheric inversions are present, is calculated. In the third group, the existence of a **triggering mechanism** is estimated by using the convective triggering temperature (Tc) and the near-surface horizontal wind divergence field. A detailed description of the aformentioned indices and typical thresholds are given in the following subsections.

2.5.1 Convective available energy and atmospheric stability

The CAPE given in J kg^{-1} (Moncrieff and Miller, 1976) is a well-known and widely used index that theoretically represents the possible intensity which convection may reach. It is defined as,

$$CAPE = \int_{z_{LFC}}^{z_{EL}} B dz = g \int_{z_{LFC}}^{z_{EL}} \frac{T_{vp} - T_{ve}}{T_{ve}} dz, \tag{2.23}$$

where z is a unit vector in the vertical, along the gravitational vector, T_{vp} is the virtual temperature of the parcel, and T_{ve} is the virtual temperature of the environment, z_{EL} is the height of the equilibrium level and z_{LFC} is the level of free convection. The CAPE must be released to produce thunderstorms. It is directly related to the maximum potential vertical speed within an updraft. In the absence of background shear and lateral entraintment, the maximum kinetic energy that can be released by a particular parcel is that which comes from converting all the potential energy into kinetic energy of vertical motion,

$$w_{max} = \sqrt{2CAPE}. \tag{2.24}$$

Higher values of CAPE indicate a greater potential for severe weather. Empirical CAPE thresholds are shown in Tab.2.2.

The LI given in °C (Galway, 1958) is a measure of the atmosphere's stability (or instability). It is a useful tool for estimation of the atmosphere's potential to produce severe thunderstorms. It is defined as,

$$LI = T_{500_{hPa,env}} - T_{500_{hPa,parcel}}, \tag{2.25}$$

where $T_{500_{hPa,env}}$ is the environmental temperature at 500 hPa and $T_{500_{hPa,parcel}}$ is the temperature at 500 hPa of a parcel lifted dry adiabatically from the 50 hPa mean of

CAPE in J kg^{-1}	Stability of the atmosphere
< 0	Stable
0 to 1000	Marginally unstable
1000 to 2500	Moderately unstable
2500 to 3500	Very unstable
$>$ 3500-4000	Extremely unstable

Table 2.2: Convective Available Potential Energy (CAPE) index thresholds

LI in °C	Stability of the atmosphere
2 or greater	stable
1 to 0	almost unstable
-1 to -4	Troposphere marginally unstable
-4 to -7	Large instability
-8 or less	Extremely instability

Table 2.3: Lifted Index (LI) thresholds

the lower PBL to the LCL and pseudo-adiabatically above. Empirical LI thresholds are Tab.2.3.

The BRN (Weisman and Klemp, 1986) is a non-dimensional ratio of the CAPE to a measure of the vertical wind shear, used to characterize convective-storm types for various environments. It is defined as,

$$BRN = \frac{CAPE}{0.5(u_{6km} - u_{500m})^2}, \tag{2.26}$$

where u_{6km} is the wind speed at 6 km above ground level (AGL), and u_{500} is the wind speed at 500 m AGL. Generally, values of the BRN lower than 45 support supercell convection, while values greater than 45 support multicell or ordinary cell convection.

BRN < 10	Strong vertical wind shear and weak CAPE. The shear may be too strong given the weak buoyancy to develop sustained convective updrafts. However, given sufficient forcing, thunderstorms may still develop.
BRN $=$ 10 to 45	Associated with supercell development
BRN > 50	Relatively weak vertical wind shear and high CAPE which suggests that multicellular thunderstorm development is most likely.

Table 2.4: Bulk Richardson Number (BRN) index thresholds

The KO-index given in K is a measure of the atmospheric potential instability (Andersson, 1989). It is computed as follows,

$$KO = 0.5(\theta_{e700} + \theta_{e500} - 2\theta_{e850}). \tag{2.27}$$

KO-index thresholds are shown in Tab.2.5,

KO > 10	Potentially very stable
6 < KO ≤ 10	Potentially stable
2 < KO ≤ 6	Potentially indifferent
KO ≤ 2	Potentially unstable

Table 2.5: KO-index thresholds.

2.5.2 Convective inhibition

The CIN given in J kg^{-1} (Colby, 1984) is the amount of energy required to overcome the negatively buoyant energy that the environment exerts on an air parcel as a result of the air parcel being cooler (more dense) than the air which surrounds it, which causes the air parcel to accelerate downward.

$$CIN = -\int_{z_{SFC}}^{z_{LFC}} Bdz = -g\int_{z_{SFC}}^{z_{LFC}} \frac{T_{vp} - T_{ve}}{T_{ve}} dz, \tag{2.28}$$

where z_{SFC} is the height near-surface or in the PBL where the parcel is lifted from. The smaller (larger) the CIN, the weaker (stronger) the amount of synoptic and especially mesoscale forced lift must be to bring the parcel to its LFC. High CIN values in the presence of little or no lift can cap or suppress convective development, despite possibly high CAPE values. Similar to CAPE, the CIN calculation also incorporates a virtual temperature correction along the parcel trajectory. Doswell and Rasmussen (1994) mention also CIN in their conclusions regarding the virtual temperature correction. They point out that CIN values are relatively small in convective situations and that an average virtual temperature change of 1 K can affect CIN by about 35 J kg^{-1} which would be significant if the CIN was approximately 100 J kg^{-1}. Using the calculated CIN, one may approximately estimate the vertical velocity, w, that an updraft will need to break through the inhibition layer,

$$w = \sqrt{2CIN}. \tag{2.29}$$

CAP describes the ability of a stable layer to inhibit low-level parcel ascent. If the CAP is strong enough, then deep moist convection will be suppressed, even if the air mass is very unstable. However, a cap allows the low-level moisture and temperature to increase which ultimately enhances severe weather potential for those stronger convective cells that are able to break the cap. Therefore, thunderstorms which develop rapidly within or near an area of significant capping are likely to become severe. Conversely, the lack of a inversion allows many storms to develop which then compete for the available moisture and storm-relative inflow. It is expressed in °C and is measured as follows,

$$CAP = MAX(T_e - T_p), \tag{2.30}$$

where T_e is the environmental temperature, T_p is the lifted-parcel temperature. CAP is calculated as the maximum temperature difference between T_e and T_p in the layer bounded between the level of the lifted parcel and the LFC. General thresholds for the CAP index are summarized in Tab.2.6.

CAP< 1°C	implies that latent and sensible heat are continuously released as small clouds, preventing sufficient energy build-up to drive strong thunderstorms.
1°C < CAP < 2°C	implies that heat and moisture trapped accumulate during the day, and serve as a fuel for future thunderstorms. This is the ideal situation.
CAP > 2°C	implies that severe thunderstorms are relatively unlikely to occur even in regions where the conditional instability is large unless sufficient heating, moisture convergence, and/or forced lift overcome the inversion.

Table 2.6: CAP thresholds

The Lid Strength Index (LSI) (Carlson et al., 1980) is a modified stability index that quantifies the convective potential of the atmosphere in the presence of lids. In order to account for the suppressive nature of mid-tropospheric inversions the LSI is defined as,

$$LSI = \underbrace{(\bar{\theta}_w - \theta_{sw5})}_{A} - \underbrace{(\theta_{swl} - \bar{\theta}_w)}_{B}, \tag{2.31}$$

where $\bar{\theta}_w$ is the vertically averaged value of θ_w in the lowest 50 hPa of the atmosphere, θ_{sw1} is the saturation wet bulb potential temperature at the warmest point in the inversion between 850 hPa and 500 hPa, and θ_{sw5} is the saturation wet bulb potential temperature at 500 hPa. Term A represents the buoyancy term and is comparable to the LI but with the sign convention for instability reversed, and it is numerically about one-half of the latter. Term B corresponds to the lid effect.

The thresholds used to interpret the results obtained for the calculation of the LSI are as follows (Tab.2.7), Graziano and Carlson (1987) also addressed the issue of a critical lid strength by means of a statistical treatment of the observations in order to develop a threshold in the use of the lid strength. They proposed that there is an effective cut-off beyond which the probability of deep convection decreases rapidly with increasing values of the lid effect term. Using the 90th percentile as an indicator of the cut-off value, the latter is found to be uniformly close to 2°C. Conditional instability, as represented by the buoyancy term (essentially the lifted index), is also important for convective outbreaks, although a critical value is less evident. Regardless of the lid effect value, the results show that the buoyancy term should be less than about 1°C for deep convection to become more than a remote possibility. Although the probability of deep convection increases with decreasing (increasingly negative) values of the buoyancy and lid terms, the probability of severe local storms, given that a thunderstorm event will occur, increases with the lid strength for a given value of the buoyancy term. Dependence of deep convection on both the lid strength and the buoyancy terms indicates that both terms should be considered jointly. Accordingly, a useful form of the lid strength concept for forecasting deep convection is a 2-D probability diagram (buoyancy term versus lid term) and the concept of a critical lid strength value, which is about 2°C.

LSI > 0	Potential exists for thunderstorms without the need for large-scale lifting.
LSI < 0	Small negative value indicates that convection may be possible locally with lifting.
	A large negative value indicates that convection will be inhibited even if the sounding has a large conditional instability.

Table 2.7: Lid Strength Index (LSI) thresholds

2.5.3 Triggering

Convection initiation in weak or absent synoptic-scale forcing is controlled by boundary layer forcing. Triggering via PBL processes can be due to thermal instabilities or dynamical mechanisms. Sometimes, the density differences at the near-surface or boundary layer, e.g. due to differential warming, allow parcels to reach their LFC. However, sometimes an additional lifting mechanism is necessary to help the parcel to overcome the inhibition present and reach its LFC. Local lifting often occurs along convergence zones, detectable by satellite (Purdom, 1982) and seen by radar as fine lines of enhanced reflectivity (Wilson and Schreiber, 1986).

Convergence can occur on different scales and be produced by different processes. Some convergence zones are due to topographical effects such as variations in terrain height or land/water boundaries (Kottmeier et al., 2008; Kalthoff et al., 2009). Others are due to variations in land-use or land-wetness, with associated variations in Bowen ratio (Weckwerth and Parsons, 2006). These spatial inhomogeneities of land use and/or soil water content result in inhomogeneities in the atmospheric temperature and moisture fields which could generate thermally induced wind systems in the boundary layer (Schaedler, 1990; Segal and Arritt, 1992) allowing the release of the convective available potential energy.

To adequatly represent and quantify the possible existence of triggering in the PBL, the analysis of the near-surface horizontal wind divergence field and the convective triggering temperature (Tc) can be used. The Tc defines the value that the near-surface virtual temperature (Tv) must reach, so that the air will reach the CCL through buoyancy and convective clouds will form, i.e., it is the temperature to which a surface parcel must be heated to allow it to freely rise. The strength of the capping inversion determines whether the Tc will be reached. When a very strong inversion exists, the Tc will be higher than the daily maximum temperature and, thus, no convection will be triggered by buoyancy. The amount of low-level moisture also determines if the Tc will be reached. The CCL height and thus Tc increase when the average PBL dewpoint decreases. Dynamic upward forcing lowers the theoretical Tc value since parcels of air can be "forced" or brought closer to the CCL. Therefore, the difference between Tc and Tv, enables us to identify the reagions where surface parcels can freely rise by buoyancy caused by differential warming so that Tc-Tv<0.

Chapter 3

Influence of Boundary Layer Processes on the Initiation of Convection

The influence of surface and atmospheric boundary layer processes on the initiation of convection is well known, especially when mid- and upper-tropospheric forcing is weak (Wilson et al., 1998). Boundary layer forcings determine the specific location where convection is triggered within larger regions of potential instability. Therefore, the question of why, where and when deep convection breaks out is highly related to boundary layer processes. Processes in this layer may generate deep convection acting as preconditioning of the environment, e.g. strongly influencing the convective indices, and/or as triggering mechanisms. In addition, PBL-capping inversions are desicive for convective inhibition. The influence of boundary layer processes on the initiation of convection can be separated into surface effects, boundary layer effects and wind circulations (Pielke, 2001).

3.1 Surface effects

The interaction between the Earth's surface and the atmosphere is critically important with respect to the development of cumulus convective rainfall. The fuel for thunderstorms is heat energy. This heat energy can be derived for example from sensible heating at the Earth's surface. Soil moisture is one of the most important factors since it affects the partitioning of the available energy into latent and sensible heat fluxes. Two aspects of surface properties can affect environmental preconditioning: 1) the actual state of the surface, dry vs. wet soil, and 2) heterogeneities in surface conditions, dry land adjacent to wet land. Wet surfaces under clear-sky conditions are generally more conducive to deep convection than dry surfaces (Segal et al., 1995). When the surface is wet, large latent heat fluxes can increase the CBL specific humidity in the afternoon, thereby enhancing CAPE. When most of the energy is transferred into sensible heat, the subsequent increase in temperature also increases CAPE. In addition, when a capping inversion is present but weak, or absent, convection can readily break out. However, in cases of stronger capping inversions, dry surfaces may be more conducive to deep convection even if there is less moisture since the larger sensible heat flux in those cases can erode the capping layer (McGinley, 1986). One may conclude that changes of conditions of the Earth's surface can result in significant changes in the energy balance. These changes will influence the heat and moisture fluxes within the PBL (Segal et al., 1989) resulting in inhomogeneities in this layer and affecting CAPE, CIN and other measures of the deep cumulus cloud activity.

3.2 Boundary layer effects

The rate of growth of the boundary layer during the day, and the entrainment of free atmospheric air into the boundary layer are both dependent on the surface heat flux. Daytime heating of the convective boundary layer (CBL) is probably the most common preconditioning process for convection over land. Typically, once the nocturnal inversion has disappeared, clouds can form as boundary layer thermals reaching their LCL. The rate at which this occurs depends on the morning inversion depth, sky cover, and surface wetness. Entrainment at the CBL top acts to dry out the boundary layer and reduce the potential for deep convection, but if moisture is sufficient, clouds can still form. Especially, boundary layer moisture inhomogeneities often have a significant impact on convection initiation (Crook, 1996; Weckwerth, 2000; Lauscaux et al., 2004). Thus, major factors determining the location of the initiation of convection are the spatial distribution and temporal development of water vapour in the CBL. In addition to evapotranspiration from soil and vegetation that makes water vapour locally available in the atmosphere, advective processes play a significant role. Direct advection of moisture into a region can increase CAPE and lower the LFC, thereby increasing the potential for deep convection. Strong moisture advection is essential for extreme precipitation events since rainout far exceeds local evapotranspiration. Moisture advection on the cloud scale may also be important for local moistening promoting new cloud growth (Perry and Hobbs, 1996). Moreover, differential advection can act on the synoptic scale, for example, in destabilizing the atmosphere (Newton, 1963), in providing the vertical wind shear (Ludlam, 1963), or in establishing capping inversions (Carlson and Ludlam, 1968). On the mesoscale, this process can result in boundary layer heating changing markedly over very short periods destabilizing relatively small regions immediately prior to convective outbreaks (Kocin et al., 1986), or in the focused removal of a capping inversion where the moist flow moves out from beneath the lid into a region of relatevily cooler air aloft, a process called underrunning (Carlson et al., 1983). If there is a strong capping inversion, the growth of clouds may be restricted or even suppressed. However, moderate capping inversions can enhance the potential for deep convection by allowing shallow cumulus to be suppressed, but then later in the day, as the inversion has been weakened by heating, a lifting mechanism can release the instability quickly (Carlson et al., 1983).

3.3 Wind systems

One of the requirements for the initiation of deep convection is the existence of a trigger mechanism like vertical motion to release the available energy for convection. As described in section 2.5.3, triggering via PBL processes can be due to thermal instabilities or dynamical mechanisms. Inhomogeneities of the boundary layer temperature and moisture fields determine if and where the convective temperature is reached in the course of the day. In these areas thermals can rise and sometimes reach their LFC. Furthermore, horizontal gradients of soil moisture or soil moisture availability often lead to secondary mesoscale circulations similar to sea-breeze flows as shown in modelling studies (Schaedler, 1990; Segal and Arritt, 1992) and observational case-studies (Taylor et al., 2007). Pielke and Segal (1986) term these thermally circulations as non-classical mesoscale circulations. The CAPE of the atmosphere is often released by regions of convergence in the CBL caused by the mesoscale circulations (Emori, 1998). Of particular importance of mesoscale circulations to deep cumulus convection is the production of focused regions of particularly favorable conditions (Pielke et al., 1991b). In these areas, CAPE and other measures of the potential for deep cumulus convection are increased in response to boundary wind convergence associated with the local wind circulations (Pielke et al., 1991a), convective inhibition is reduced and the wind convergence zones can also provide specific vertical motion to initiate deep cumulus convection. Thus, surface inhomogeneities in soil moisture or vegetation type can lead to both preconditioning and triggering of convection (Anthes, 1984).

Terrain forcing is also a common trigger of convection. Various ways in which this can occur are discussed by Houze (1993). Small-scale topographic features can contribute directly to the initiation of severe convective weather through localized thermally induced lifting or earlier attainment of the convective temperature. Surface relief, whether it be small hills, ridges, escarpments, or mountain ranges, can have profound effects on convection. Banta (1990) identified three classifications of topographic effects on convection: mechanical lifting to the LFC, directly related to triggering, thermally generated circulations, and aerodynamic effects such as blocking, flow deflection etc.. Thermally generated flows such as up-slope and up-valley wind systems can play a prominent role in the initiation and development of severe weather events.

Convergence zones of different origin are other important PBL phenomena which initiate deep convection (Fankhauser et al., 1985; Wilson and Schreiber, 1986; Wilson et al., 1992, 1998; Kottmeier et al., 2008; Kalthoff et al., 2009). Convection often initiates along sharp PBL convergence lines. But, shallow convergence lines, inadequate to trigger convection, may deepen the moist layer such that subsequent or additional lifting is more effective at initiating convection. Thus, convergence lines can serve as both preconditioning and triggering mechanisms. Localized convergence and lifting destabilize the environment and reduce CIN, thereby making the atmosphere susceptible to deep convection. In some cases, convergence can be so strong that deep or severe convection can occur even in the absence of high CAPE (Carbone, 1982). Mountain/valley breezes are also important for preparing the atmosphere for severe weather (Maddox et al., 1980), by advection of warm moist air and/or creating convergence lines along the top of the mountain crests (Meissner et al., 2007).

Low-level convergence also occurs at the leading edges of sea and land breezes. Sea breeze circulations are driven by horizontal pressure gradients resulting from daytime differences in temperature between cool air over water and warm air over land (Fig.3.1). The cool marine air associated with the sea breezes generally acts to suppress convective activity over water and at inland locations. However, the leading edge of the sea breeze circulation, called the *sea breeze front*, is a zone of en-

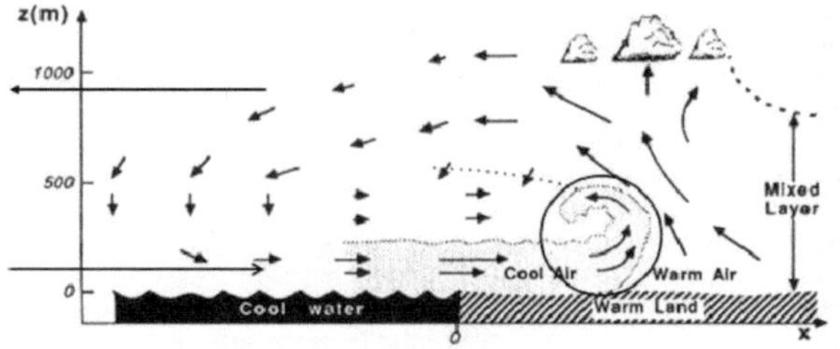

Figure 3.1: Schematic sketch of a typical sea breeze front (After Lyons (1975) and Ogawa et al. (1986)).

hanced moisture and lift that can act to trigger the development of thunderstorms in a convectively unstable environment. This is especially true when sea breeze fronts interact with thunderstorm gust fronts, synoptic-scale fronts, or other sea breeze fronts produced by differently oriented coastlines. The arrival of the sea breeze front is frequently associated with a rapid decrease in temperature, a rapid increase in relative humidity, and a shift to onshore winds. However, the arrival signature in the temperature and relative humidity data typically becomes more subtle with increasing inland distance due to the rapid modification of the marine air mass over land. A large number of studies in other areas linked sea breezes with severe convection showing that convective thunderstorms frequently form near and ahead of sea breeze fronts. In Florida, there is a long history of studies relating sea breezes and convection (Byers and Rodebush, 1948; Pielke, 1974; Kingsmill, 1995; Wilson and Megenhardt, 1997). In Australia, the Mesoscale Convective Systems (MCSs) that regularly form over the Tiwi Islands originate from convection initiated by sea breezes (Carbone et al., 2000). Coastline shape, synoptic scale winds, and water bodies close to the coastline influence sea breeze frontal morphology (Laird et al., 1995). Moreover, numerical studies (Zhong et al., 1991) have shown that the convergence patterns (i.e., vertical motion fields) associated with the sea breeze are altered along irregular coastlines. Even lake breezes can contribute to the formation of severe thunderstorms and tornados (King and Sills, 1998). In some cases, the low-level lifting provided by sea, land, or lake breezes is adequate by itself to generate convection along the boundary. These phenomena sometimes do not trigger convection by their own. However, it is the intersection or collision of these phenomena with each other, or with other features such as terrain, that provides the most energetic triggering of convection (Wilson and Schreiber, 1986).

Chapter 4

The CSIP and COPS Projects

Damage caused by heavy rain is a problem that is motivating research as a major focus for international activities at the present time. Thunderstorms and rain associated with deep convection are highly localized events which are poorly forecasted. A key task facing the meteorological community is to gain a better understanding of why deep convection breaks out precisely where and when it does and to use such an understanding to improve NWP models. These goals played a major role in three complementary field campaigns conducted in the last years. One of these, known as the International H_2O Project (IHOP_2002; Weckwerth et al., 2004), took place during the summer of 2002 in the U.S. southern Great Plains, a region characterized by large convective instability and strong capping inversions. The local orographic variations in this region are small and generally not critical to the triggering of convection. Another experiment was the Convective Storm Initiation Project (CSIP) (Browning et al., 2007). The CSIP field campaigns during the summers of 2004 and 2005 took place in the southern region of the United Kingdom (UK), which is characterized by an intermediate level of orography and nearby coastlines. It is due to the mainly maritime nature of the British climate and the remoteness of any major mountainous areas that the convective instability and capping inversions are often quite weak on convective occasions. The last experiment, the Convective and Orographically-Induced Precipitation Study (COPS) (Wulfmeyer et al., 2008; Kottmeier et al., 2008), during summer 2007, took place in southwestern Germany and eastern France. This region is also characterized by large instability but with a major orographic influence.

In this thesis, data from the CSIP and COPS experiments are used to gain a better understanding of the initiation of deep moist convection in Europe, in a maritime almost flat region such as the southern UK (CSIP), and in a topographically complex terrain such as southwestern Germany (COPS). In this chapter, a general overview of both measurement campaigns is presented.

4.1 Convective Storm Initiation Project (CSIP)

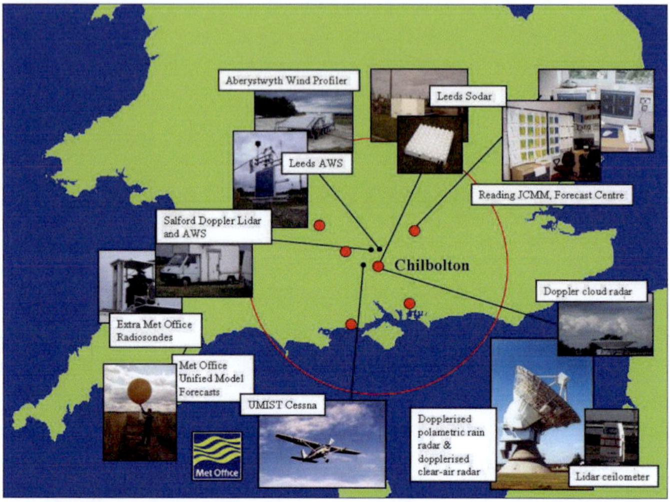

Figure 4.1: Map showing locations of instruments deployed in southern Britain during CSIP. Positions of the radiosonde stations (red circles), from left to right; Bath(B), Swanage(S), Larkhill(L), Chilbolton(C), Reading(R), and Preston Farm(P) (Adapted from Browning et al. 2007).

The CSIP main field campaign (Browning et al., 2007) was conducted over southern England in June, July and August 2005. An earlier pilot campaign was performed in the same region in July 2004. An overview of the observational setup is given in Fig.4.1. The operation center was based at Chilbolton on the premises of the Chilbolton radar facility. Other instruments were sited within the range of the Chilbolton radar. Within this range six radiosonde stations are located; Bath(B), Swanage(S), Larkhill(L), Chilbolton(C), Reading(R), and Preston Farm(P), each separated from its nearest neighbour by 25 to 50 km. During the Intensive Observation Periods (IOPs) the radiosondes were launched every one or two hours. A network of Automatic Weather Stations (AWSs) was located around Chilbolton. Mobile synoptic observation stations and a network of 5 GeoForschungsZentrum (GFZ) Potsdam Global Positioning System (GPS) integrated water vapour stations in addition to the GPS water vapour stations used by the Met Office and the ones from the University of Bath were available. Additionally, surface data from SYNOP stations, energy balance stations located at Chilbolton and Bath for some of the IOPs, and 2 instrumented light aircrafts (Dornier-128 and Cessna 182) among other instruments were available. The UK operational radar network provided maps of estimated rainfall intensity at resolutions mainly between 1 and 2 km every 5 min for single radars and every 15 min for composite displays. High-resolution visible imagery every 15 min was available from Meteosat-8, the Meteosat Second Generation (MSG) satellite.

The southern region of the UK is characterized by an intermediate level of orography (Fig.4.2). Within the domain limited by the six radiosonde stations in the investigation area, the highest elevations are found in the north-western region with hills up to about 180 m AGL. Despite the low-level elevations it has been proven that the topography of the UK has some influence on the initiation of convection (Bennett et al., 2006). Eighteen IOPs took place during the three-month period of June, July and August of 2005. In all IOPs, except one, convection was initiated in the boundary layer (Browning et al., 2007). A detailed description of the experimental setup and the main goals of the CSIP campaign can be found in Browning et al. (2007).

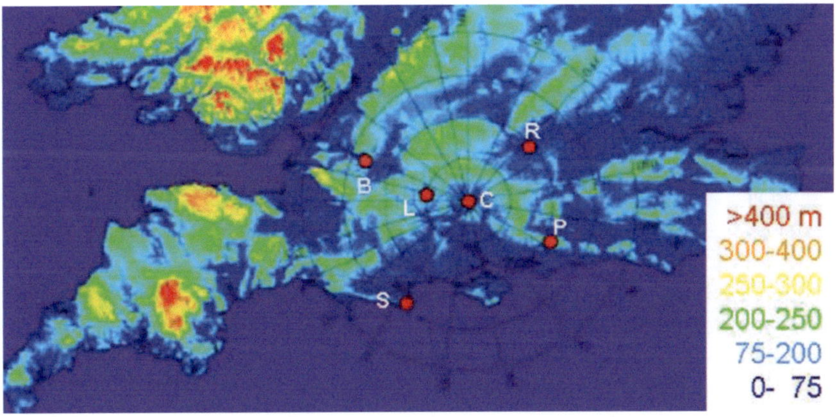

Figure 4.2: Topography of the CSIP domain. The positions of the radiosonde sites are indicated by full red circles. Range rings centered on the Chilbolton radar are shown at 25 km intervals. (Adapted from CSIP IOP Summary, Browning and Morcrette, 2005).

4.2 Convective and Orographically-Induced Precipitation Study (COPS)

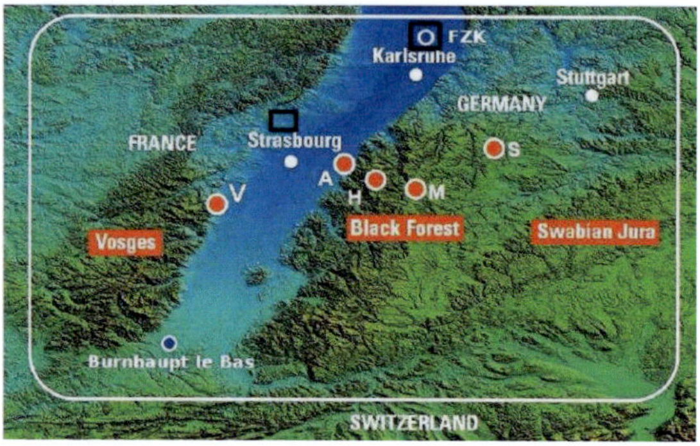

Figure 4.3: The COPS domain in southwestern Germany and eastern France. Supersites (orange circles), from left to right; V(Vosges/Meistratzheim), A(Achern/Rhine valley), H(Hornisgrinde), M(Murgtal/Heselbach), and S(Stuttgart/Deckenpfronn). Additional radiosonde stations (blue circles), Burnhaupt le Bas and Forschungszentrum Karlsruhe (FZK). The locations of precipitation radars to the north of Strasbourg and FZK northward of Karlsruhe are marked with black squares (Kottmeier et al., 2008).

The COPS field campaign (Wulfmeyer et al., 2008; Kottmeier et al., 2008) was performed from 1 June to 31 August 2007 in southwestern Germany and eastern France. Five supersites with a large variety of different instruments were deployed in the COPS region. The supersites were called supersite V (Vosges, low mountain region), R (Rhine valley), H (Hornisgrinde, mountain site), M (Murg valley), and S (Dornstetten close to Stuttgart)(Fig.4.3). They were equipped each with a radiosonde and standard meteorological measurements as well as different remote sensing instruments (e.g. wind and water vapour lidar, cloud radar, radiometers, microwave radar) resulting in a beneficial sensor synergy. Two additional radiosonde stations in Burnhaupt le Bas and at Forschungszentrum Karlsruhe (FZK) provided additional information about the vertical structure of the atmosphere at the northern and southern borders of the COPS region. The IMK C-band radar at FZK and the C-band Poldirad of DLR Oberpfaffenhofen located close to Strasbourg extended the existing German, French and Swiss radar networks. Research networks measuring soil moisture, energy balance, near-surface turbulence or standard meteorological parameters were set up, and the GPS station network was densified. A sodar and wind profiler network was operated to study the boundary layer wind field. Rapid Scanning (RS) service from Meteosat-8 (MSG) provided 5 min scans which were used to observe the evolution of convective clouds and to detect convection initiation sites (Aoshima et al., 2008). Altogether 9 aircraft and an instrumented airship Zeppelin NT were operated, either dedicated to boundary layer observations (e.g. DO 128, Dimona motoglider) or mapping the large-scale conditions (e.g. German and French Falcons F20, British BAe146). The operation center was located at Baden-Airpark in the Rhine valley. During the COPS field campaign, 18 IOPs were conducted providing a comprehensive data set covering many different atmospheric conditions. Due to the fact that some IOPs lasted for 2 or more consecutive days, the number of IOP days was 35. The individual days regarding the type of observed convection were classified by Kottmeier et al. (2008) as: (i) locally initiated convection, (ii) widespread convection triggered by large-scale lifting of potentially unstable air masses, and (iii) embedded convection near convergence lines or frontal zones. However, a mixture or combination of these categories was necessary to describe those days, which could not be categorized into one single category. The COPS domain can be identified as a complex terrain (Fig.4.4). Topography in the area ranging from 100 m to more than 1 km AGL have a significant influence on the initiation of convection in the area.

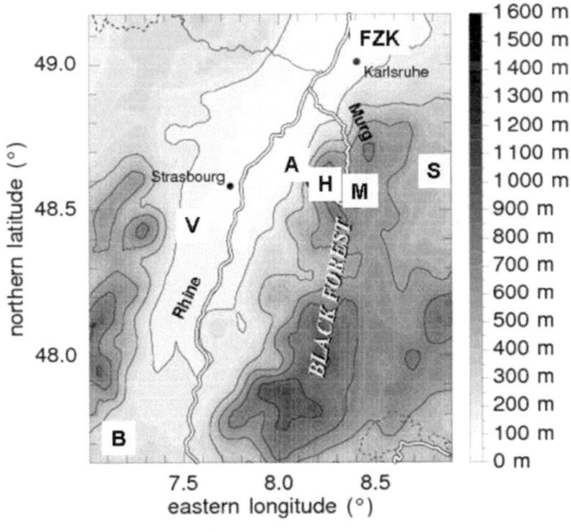

Figure 4.4: Topography of the COPS domain in the northern Black Forest. Supersites and radiosonde stations as in Fig.4.3.

Chapter 5

The Spatial Variability in the Boundary Layer and Its Impact on Convection Initiation

Radiosondes remain an essential tool for measuring boundary layer thermodynamic quantities. However, the coarse spatial resolution of radiosoundings may miss the potentially significant variability induced by surface and boundary layer processes (Weckwerth et al., 1996), and a single sounding alone will typically not be useful for evaluating the potential for thunderstorm initiation (Weckwerth, 2000). Nevertheless, radiosondes have a high vertical resolution, capable of defining the bases and thicknesses of cloud layers and stable layers. They also give the only operationally available upper-air data with an absolute calibration (Lorenc et al., 1996).

In this chapter it is investigated whether the dense CSIP network of radiosondes, of about 25 to 50 km (Fig.4.1), especially deployed for CSIP, is able to resolve near-surface and boundary layer variability expected to be relevant for the initiation of convection. The spatial variability of the meteorological parameters below the resolution of the radiosonde network and its impact on the convection indices is then analyzed by aircraft measurements. To illustrate the analysis, data from the CSIP IOP5, 29 June 2005, is used. On this day, although deep convection resulting in heavy rain and flash flooding was observed out of the CSIP area of investigation, the initiation of deep convection did occur in the north-eastern region of the CSIP domain, between Chilbolton and Reading radiosonde stations (Fig.4.1), in the period from 1000 to 1200 UTC. The convection activity and synoptic situation observed in this day are explained more in detail in Chapter 7.

5.1 The impact of spatial variability in the PBL observed by the radiosonde network on convection-related indices

In this section, the measurements from radiosoundings, with a spatial resolution of approximately 25 to 50 km, are used to investigate the boundary layer moisture and temperature spatio-temporal variability. Convection indices are calculated to determine the atmospheric conditions at the radiosonde sites.

5.1.1 Moisture and temperature in the boundary layer using radiosonde profiles

The radiosonde profiles at the six radiosonde stations are analyzed in the period from 1000 to 1200 UTC, time period when the initiation of convection was observed. As an example, the profiles at Reading and Chilbolton up to 3 km amsl are presented in Fig.5.1. At 1000 UTC there are already significant variations in the PBL's temperature and moisture (Tab.5.1). The maximum differences in temperature are about 2.5 °C and in humidity about 1.5 g kg^{-1}. At that time, Reading shows the coldest boundary layer, 14.6 °C, and the lower boundary layer at Bath is almost 1 g kg^{-1} drier than those at the rest of the stations. At 1200 UTC, the maximum differences in temperature are about 3 °C while the differences in humidity are about 2 g kg^{-1}. In the period from 1000 to 1200 UTC, the boundary layer at Reading is observed to warm more rapidly than at the other sites (Fig.5.1), probably because this site was less affected by the cirrus cover than the other sites as suggested by Marsham et al. (2007b). In that station, a moisture increase in the order of 0.5 g kg^{-1} is observed in the time period evaluated whereas in the rest of the stations almost no change or drying occurs. The spatial and temporal temperature and moisture variations observed in the boundary layer caused considerable spatial differences in the convection indices, which are significant for convective initiation (Tab.5.2).
By 1000 UTC, very low CAPE below 100 J kg^{-1} and positive LI in almost all stations, indicate that the atmosphere at this time is quite stable. At Reading, the calculated value of CIN is approximately -150 J kg^{-1}. By 1200 UTC, CAPE in the area ranges from low conditional instability at Swanage (23 J kg^{-1}) to moderate values at Reading (555 J kg^{-1}). The little increase in temperature and the drying observed in most of the stations (Tab.5.1) leads to a small increase in CAPE at these stations. Only in the areas of Preston and Reading, with the highest temperature and moisture in the lowest PBL at this time (Tab.5.1), a significant atmospheric destabilization occurs. In addition, at Reading the altitude of the LFC is reduced

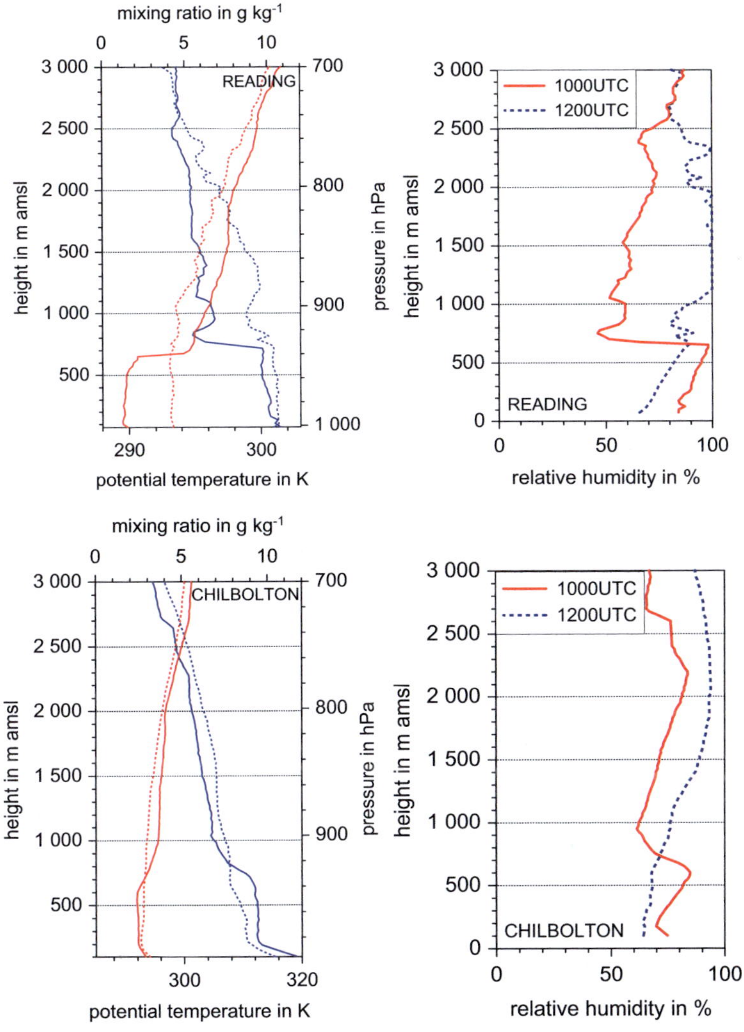

Figure 5.1: Potential temperature (red full line at 1000 UTC and red dotted line at 1200 UTC), mixing ratio (blue full line at 1000 UTC and blue dotted line at 1200 UTC) (left) and relative humidity (RH) (right) at Reading and Chilbolton on 29 June 2005.

	Time	Bath	Larkhill	Chilbolton	Reading	Swanage	Preston F.
T in °C	1000	15.2	16.0	16.6	14.6	14.9	17.0
T in °C	1200	16.7	16.6	17.3	18.1	15.0	17.1
r in g kg^{-1}	1000	8.1	9.3	9.2	9.5	9.1	9.7
r in g kg^{-1}	1200	7.9	8.4	8.5	10.0	9.1	9.9

Table 5.1: Temperature (T) and mixing ratio (r) mean of the lower 50 hPa in the PBL at the six radiosonde stations in the CSIP domain, at 1000 and 1200 UTC on 29 June 2005.

	LCL in hPa	LFC in hPa	CAPE in J kg^{-1}	CIN in J kg^{-1}	LI in °C	CAP in °C
1000 UTC						
Bath	882	838	0	-27	2.4	1.8
Larkhill	920	915	1	-2	-0.1	0.2
Chilbolton	905	805	7	-62	0.6	2.9
Reading	953	726	0	-147	1.9	3.8
Swanage	936	902	7	-14	0.8	1.2
Preston Farm	925	833	62	-43	-0.1	2.1
1200 UTC						
Bath	868	831	152	-16	1.1	1.3
Larkhill	888	859	67	-8	0.2	0.7
Chilbolton	884	835	146	-21	0.5	1.1
Reading	911	903	555	-2	-2.3	0.4
Swanage	942	843	23	-10	0.3	0.7
Preston Farm	928	887	425	-11	-1.4	1.1

Table 5.2: Lifting condensation level (LCL), level of free convection (LFC), convective available potential energy (CAPE), convective inhibition (CIN), lifted index (LI), and cap strength (CAP), at the radiosonde sites at 1000 UTC and 1200 UTC on 29 June 2005. The parameters are calculated using the 50 hPa ML parcel choice.

significantly, from 726 hPa at 1000 UTC to 903 hPa at 1200 UTC. Inhibition has decreased at all stations. CIN is very low in the whole area ($>$ -10 J kg^{-1}). The decrease of inhibition observed at Reading from 1000 UTC (CIN $\sim$ -150 J kg^{-1}) to 1200 UTC (CIN $\sim$ -2 J kg^{-1}) due to the strong increase in PBL temperature and moisture deserves a special mention. If the PBL's mixing ratio from Reading was used at Chilbolton, while mantaining all the other variables, the CAPE would increase to 800 J kg^{-1} and the CIN would be about 0.5 J kg^{-1}. This highlights the sensitivity of the convection indices to the boundary layer moisture content, thus the interpretation of the atmospheric conditions one obtains from them. The base and top of a cloud layer can be defined by identifying the level at which the relative humidity reaches maximum values, close to 100 %, with lower values below and above, respectively. Within the cloud layer, the mixing ratio decreases slightly with height. At Reading, at 1000 UTC, a cloud-topped PBL is observed (Fig.5.1). At 1200 UTC, a saturated cloud layer extends from $\sim$ 1 to 2 km amsl (Fig.5.1). At that time, light precipitation is observed in this area.

5.1.2 Local temperature and humidity change as predictors for the evolution of the convection indices

To identify and quantify the processes responsible for the temperature and moisture variations in the boundary layer, heat and moisture budget equations are calculated at Chilbolton using a simplified form of the Reynolds averaged prognostic equations for mean potential temperature, θ, and specific humidity, q (Stull, 1988). According to these simplified equations, the net local change of temperature/moisture, Lc, is produced by horizontal advection in the x-direction, Ax, advection in y-direction, Ay, and turbulent flux divergence, Fd (Khodayar et al., 2008). A residuum term is calculated as, Res = Lc - (Ax+Ay+Fd), representing the contribution of all the terms that have been neglected or cannot be calculated. Budget equations were calculated for Chilbolton since for this station surface energy balance components are available.

The integration over the depth of the PBL, $z_i \sim 600$ m AGL, of the local change of θ and q shows that warming in the order of 232 K m h^{-1} and drying in the order of -247 g kg^{-1} m h^{-1} are observed, respectively. Unfortunately, the very large residuum terms obtained for the heat and moisture budgets, $z_i\ Q_{res} \sim 455$ K m h^{-1} and $z_i\ q_{res} \sim -465$ g kg^{-1} m h^{-1}, respectively, indicate that the bundle of uncertainties and assumptions made result in a residuum of the same order of magnitude of the rate of heat/moisture atmospheric storage. This suggests that the PBL drying and warming observed at Chilbolton cannot be explained by a combination of advection and turbulent flux divergence, thus other processes such as vertical advection or diabatic processes, the latter including radiative flux convergence or divergence and phase changes, may dominate the atmospheric evolution. Due to the significant error in this calculation the results are not shown. This demonstrates that the calculation of heat and moisture budgets is limited to the availability of measurements and it is not always possible to identify the processes responsible for the temporal variations of temperature and moisture in the PBL.

However, when radiosondes are available, local variations in time, Lc, can be easily calculated and used as a predictor for the evolution of the atmospheric conditions. Tab.5.3 and Fig.5.2 show the local mean PBL potential temperature change ($\Delta\bar{\Theta}/\Delta t$) and humidity change ($\Delta\bar{r}/\Delta t$) between 1000 and 1200 UTC at the radiosonde sites. In agreement with the observations of the lower boundary layer, (Fig.5.1), at most of the stations little warming and drying is observed in the PBL whereas Reading is the station where the maximum warming, $\Delta\bar{\Theta}/\Delta t \sim 1.7$ K h^{-1}, and moistening, $\Delta\bar{r}/\Delta t \sim 0.3$ g kg^{-1} h^{-1}, effect is observed. As discussed in the last section, a temperature and/or moisture increase in the PBL, while the conditions above remain the same, implies an increase in atmospheric instability as well as a decrease in inhibition, therefore, favoring the occurrence of deep convection in the area. Based on this fact, one may expect Reading to be the area where deep convection is more likely. This is in agreement with the observations.

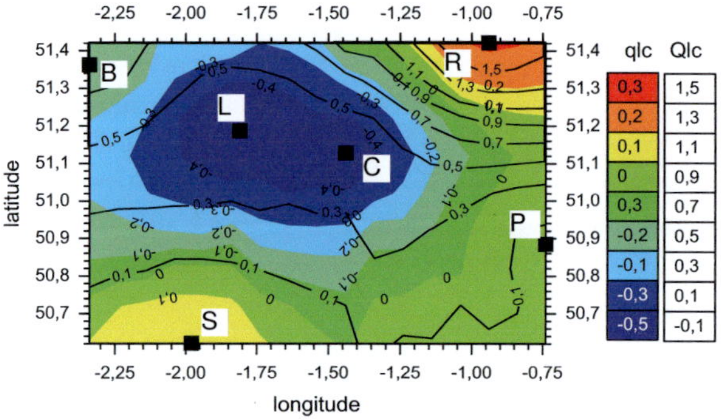

Figure 5.2: Spatial representation of the local mean PBL potential temperature change, $\Delta\bar{\Theta}/\Delta t$ in K h^{-1} (isolines), and humidity change $\Delta\bar{r}/\Delta t$ in g kg^{-1} h^{-1} (colour code), between 1000 and 1200 UTC by the interpolation of the observations at the radiosonde sites.

	Bath	Larkhill	Chilbolton	Reading	Swanage	Preston
$\Delta\bar{\Theta}/\Delta t$ in K h^{-1}	0.7	0.3	0.4	1.7	-0.1	0.1
$\Delta\bar{r}$ /Δt in g kg^{-1} h^{-1}	-0.1	-0.5	-0.5	0.3	0.2	0.1

Table 5.3: Local mean PBL potential temperature change ($\Delta\bar{\Theta}/\Delta t$) and humidity change ($\Delta\bar{r}/\Delta t$) between 1000 and 1200 UTC at the radiosonde sites.

5.2 The impact of small-scale temperature and moisture variability in the boundary layer on convection indices

To investigate the spatial variability of the meteorological parameters below the resolution of the radiosonde network and its impact on the convection indices, aircraft measurements are analyzed.

5.2.1 Temperature and moisture variability in the PBL using aircraft measurements

Two flights took place during CSIP IOP5. The first, by an instrumented Dornier-128 (Do-128), from 0900 to 1238 UTC towards the west of Chilbolton, and the second, by an instrumented Cessna, from 1029 to 1148 UTC towards the northeast of Chilbolton. Temperature, water vapour (observed using a Lyman-alpha system and a HUMICAP system), pressure, radar height above ground, and radiative fluxes are available at very high resolution from instruments on the Do-128. Temperature, water vapour, pressure, radar height above ground, and upwelling long-wave irradiance were measured at 1 Hz by instruments on the Cessna. The Cessna flight was at an altitude of approximately 700 m, almost all of it under moderate cirrus cover. When the Do-128 was at 420 m before 1041 UTC, it flew under moderate cirrus and in clear-sky regions. After 1046 UTC, it flew at 510 m, under thick and moderate cirrus and clear skies (Marsham et al., 2007a).

To combine radiosonde and aircraft measurements, the possible systematic differences in temperature and humidity between both systems were estimated. No difference for the temperature, but a bias of 1.5 ± 0.4 g kg^{-1} between the Do-128 mixing ratio data and the radiosonde data was found. This was determined from several comparisons between flight tracks over the Chilbolton and Bath radiosonde sites and the nearest radiosonde in time (i.e. within 33 min) over the corresponding height interval. The bias was removed from the aircraft data. The Cessna data agreed well with the radiosonde data on this day. Additionally, a running mean with an averaging length of approximately 1 km was used to reduce turbulence effect in the time series on the variability, which was assumed to be relevant to convection initiation. The data are only used to analyze horizontal variations in moisture and temperature. Fig.5.4 and Fig.5.3 show the temperature in °C and mixing ratio in g kg^{-1} measurements from the Cessna and the Do-128 flight tracks, respectively. In addition, the KT15 measurements of the surface (skin) temperature are shown (Fig.5.5). At both levels, $\sim$ 500 m amsl (Do-128) and 700 m amsl (Cessna), moisture shows higher spatial variability whereas temperature distribution is more homogeneous. While temperature variability is somewhat reduced in the upper PBL compared to the near-surface, moisture variability is still significant (Figs.5.3, 5.4, 5.5). The spatial variability of temperature and moisture is investigated using data from the Do-128 flight along a 36 km track, approximately 10 min flight time, in the area of Chilbolton at about 510 m amsl, from point 5 to point 1 (Fig.5.3a), at about 1200 UTC (Tab.5.4). The temperature shows a variation of about 1.5 °C and the mixing ratio of about 2.5 g kg^{-1}, which is in the order of magnitude as the variability shown between the radiosonde sites at this time (Tab.5.1). The surface temperature shows a variability in the order of 9 to 9.5 °C.

	Minimum	Radiosonde	Maximum	σ
T in °C	14.5	15.5	16.0	0.3
r in g kg^{-1}	6.9	8.2	9.5	0.5
KT15 T in °C	15.4	21.3	24.7	1.7

Table 5.4: Minimum, maximum and standard deviation (σ) of temperature (T), and mixing ratio (r) and KT15 surface temperature from the 36 km Do-128 flight track over the Chilbolton radiosonde site, as well as the radiosonde measurements at the corresponding level.

5.2.2 Impact on the convective indices

The temperature and moisture variations measured along the 36 km flight track are used to evaluate the effect that boundary layer variability has on convection indices (Weckwerth, 2000). These spatial variations were used to modify the radiosonde profiles at Chilbolton, at 1200 UTC, introducing the temperature and mixing ratio variations (minimum, maximum, and σ) at the same level as the aircraft was flying (Fig.5.6). The convection indices are calculated lifting a parcel from the flight level (Tab.5.5).

Using the maximum and minimum temperature and moisture, a difference of approximately 680 J kg^{-1} in CAPE, $\sim$ -150 J kg^{-1} in CIN, $\sim$ 5.1 °C in LI, and $\sim$ 2 °C in CAP is found (Tab.5.5). When using the standard deviations, the atmospheric conditions vary from low probability (CAPE = 70 J kg^{-1} and CIN = -36 J kg^{-1}) to moderate probability (CAPE = 302 J kg^{-1} and CIN = -1 J kg^{-1}) of deep convection. This means, that the variability of the convection-related indices below the resolu-

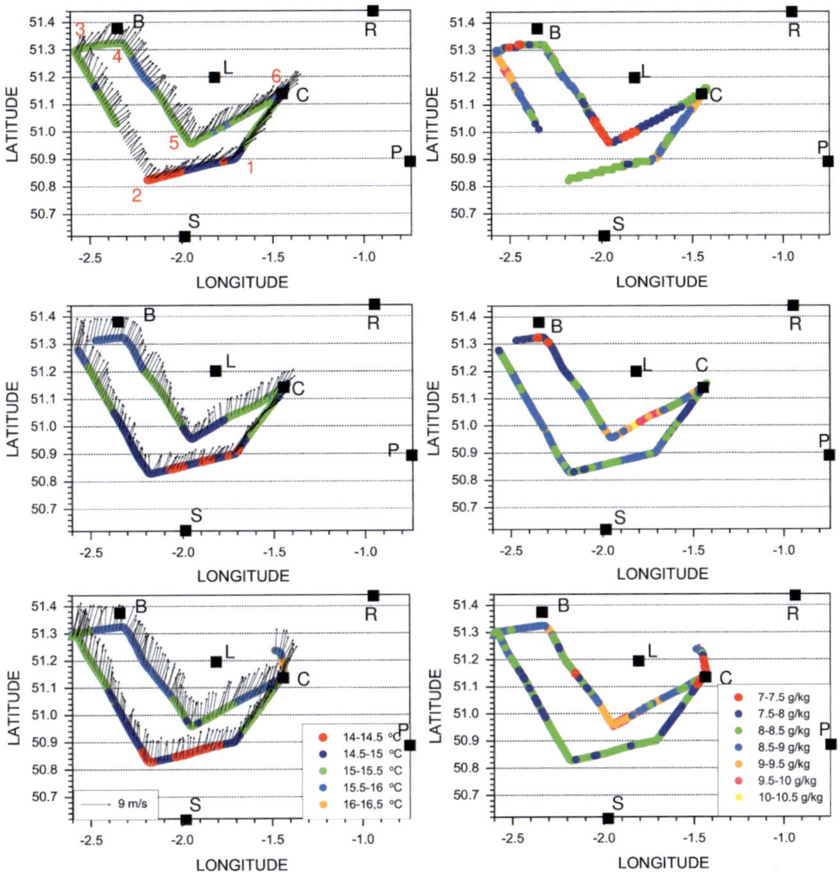

Figure 5.3: Spatial distribution of temperature in °C (left) and mixing ratio in g kg^{-1} (right) from Do-128 measurements in the time periods: (top) from 1022 to 1126 UTC, (center) from 1126 to 1230 UTC, and (bottom) from 1324 to 1429 UTC. In all periods, the flight starts at point 6 toward point 1, finishing again at point 6. The letters indicate the positions of the radiosonde stations. The arrows indicate wind vectors.

tion of the radiosondes is of the same order of magnitude as for the resolution of the radiosondes, thus an enhanced resolution for predicting the areas of initiation of deep convection would be helpful. Moreover, the PBL variability and its impact on the convection indices is investigated along a 11 km track (Tabs.5.6, 5.7). The

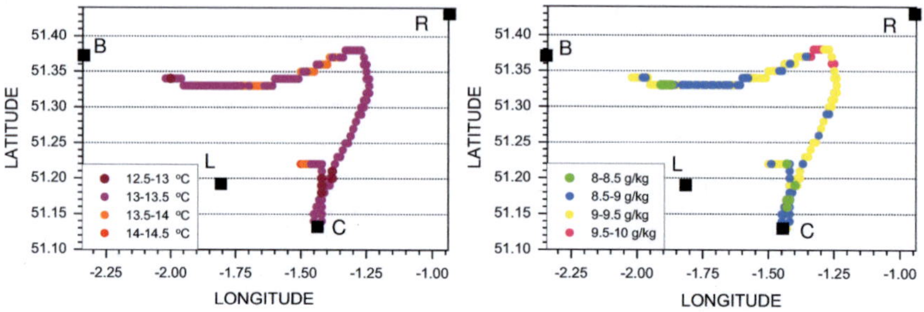

Figure 5.4: Temperature in °C (left) and mixing ratio in g kg^{-1} (right) spatial distribution from Cessna measurements for the time period from 1324 to 1429 UTC. The position of the radiosonde stations are indicated by black squares.

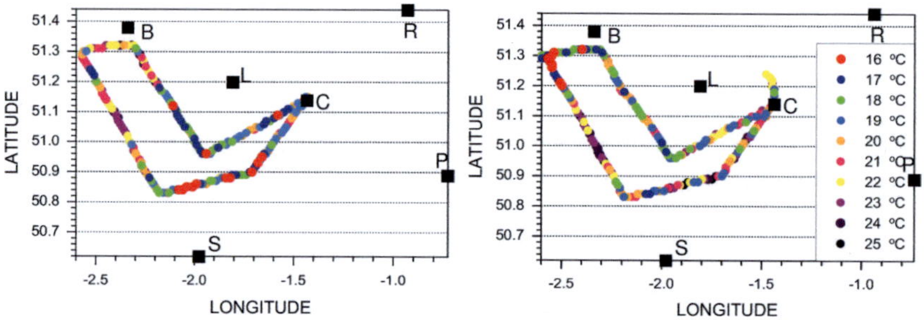

Figure 5.5: KT15 surface temperature measurements from the Do-128 in the time period from 1022 to 1126 UTC (left) and from 1324 to 1429 UTC (left) on 29 June 2005.

temperature variation measured along the 11 km track is almost negligible, however, moisture spatial distribution still shows a variation of nearly 1 g kg^{-1}. The use of the maximum/minimum temperature and moisture in the radiosonde profiles or the standard deviations of these quantities shows that low probability for deep convection is obtained in all cases.

From this analysis one may conclude that even the dense net of radiosondes deployed for the CSIP campaign is not enough to fully represent the boundary layer variability relevant to the initiation of deep convection. Therefore, an increase in resolution of the PBL temperature and moisture fields is necessary to adequately diagnose the likelihood of deep convection. A resolution in the order of 11 km would reduce significantly the uncertainty in the prediction.

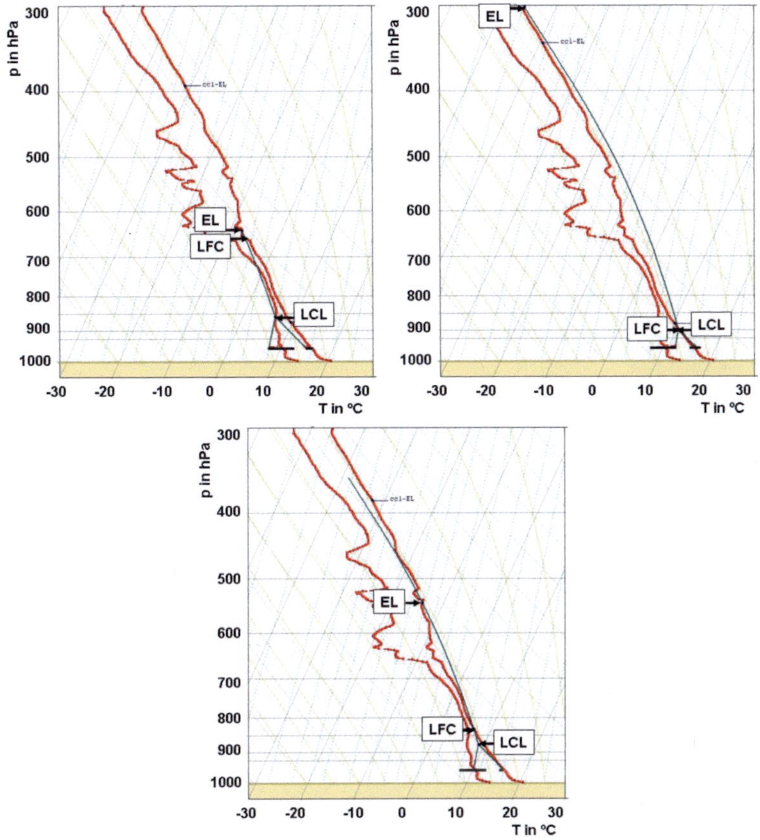

Figure 5.6: Radiosonde profile (red lines) at Chilbolton on 29 June, 2005 at 1200 UTC with lifted condensation level (LCL), level of free convection (LFC) and equilibrium level (EL) based on a parcel lifted (blue line) from the minimum (top-left), maximum (top-right) values measured along a 36 flight track at 510 m amsl over Chilbolton, as well as from the radiosonde data measured at the same level (bottom).

	$(T,r)_{min}$	$(T,r)_{rad}$	$(T,r)_{max}$	$(T,r)_{rad} + \sigma$	$(T,r)_{rad} - \sigma$
CAPE	4	156	684	302	70
CIN	-150	-11	0	-1	-36
LI	3.4	0.7	-1.7	-0.2	1.6
CAP	3.3	2.1	1.3	1.8	2.5

Table 5.5: CAPE, LI, CIN, CAP at the Chilbolton radiosonde station at 1200 UTC on 29 June 2005. The convection indices are based on minimum (T, r min) and maximum temperature and moisture (T,r max) measured at the flight track, radiosonde temperature and moisture at the same level (T,r rad), radiosonde temperature/moisture plus (T,r rad+σ) and minus (T,r rad-σ) the standard deviation of the flight track.

	Minimum	Radiosonde	Maximum	σ
T in °C	15.5	15.5	15.9	0.1
r in g kg^{-1}	7.6	8.2	8.5	0.2

Table 5.6: Minimum, maximum and standard deviation (σ), of temperature (T) and mixing ratio (r) from the 11 km Do-128 flight track over the Chilbolton radiosonde site, and radiosonde data at the same level.

	$(T,r)_{min}$	$(T,r)_{rad}$	$(T,r)_{max}$	$(T,r)_{rad} + \sigma$	$(T,r)_{rad} - \sigma$
CAPE	67	156	283	193	128
CIN	-32	-11	-1	-7	-15
LI	1.7	0.7	-0.1	0.4	1.0
CAP	2.3	2.1	1.7	2.0	2.2

Table 5.7: CAPE, LI, CIN, CAP at the Chilbolton radiosonde station at 1200 UTC on 29 June 2005. The convection indices are based on minimum (T, r min) and maximum temperature and moisture (T,r max) measured at the 11 km flight track, radiosonde temperature and moisture at the same level (T,r rad), radiosonde temperature/moisture plus (T,r rad+σ) and minus (T,r rad-σ) the standard deviation of the flight track.

Aircraft measurements although useful for evaluating the PBL variability are expensive and rarely available, and the horizontal coverage of radiosondes is low because of high costs. The Global Positioning System (GPS) technology (Wickert and Gendt, 2006; Ware et al., 1996; Bevis et al., 1992) has great potential for this application. This will be shown in the next chapter.

Chapter 6

Increase of the Spatial data Resolution by Near-Surface measurements and GPS Integrated Water Vapour data

In the previous chapter, it was demonstrated using aircraft observations within the CBL that important moisture and temperature variations exist below the resolution of the radiosonde network which have a significant influence on convection indices, thus, on the atmospheric representation obtained from them. It was therefore suggested, in agreement with former studies (Weckwerth et al., 1996), that the typical resolution of the radiosonde and surface network is not sufficient to sample CBL variability. Not even in the CSIP campaign, where a dense net of radiosondes was deployed.

Some very promising technologies are being investigated to obtain a better representation of the water vapour field. For example, the combination of ground-based lidar systems to measure turbulent fluxes has proved to be a suitable tool for the investigation of transport processes within the boundary layer (Linne et al., 2007). Also the use of scanning water vapour differential absorption lidar to obtain a three-dimensional mapping of the water vapour field shows a great deal of promise for research applications (Wulfmeyer, 1998). Ground-based water vapour radiometers (WVRs) are instruments that scan the sky and measure the microwave radiation emitted by atmospheric water vapour. The frequency dependence of sky brightness temperature enables the simultaneous estimation of IWV and integrated liquid water along each line of sight in the scanning pattern. Most meteorologists are more familiar with space-based, downward looking WVRs. Satellite-based WVRs provide good spatial coverage but poor coverage in time, whereas ground-based WVRs have the opposite characteristics (Bevis et al., 1992).

The prospect of using earth-based GPS technology (Wickert and Gendt, 2006; Ware et al., 1996; Bevis et al., 1992) to measure atmospheric water vapour for research and operational weather forecasting is promising because extensive networks of continuously operating GPS receivers are being installed around the world. GPS-derived IWV data could be made available to the research and operational meteorological community at relatively little incremental cost. Moreover, the GPS IWV measurements provide continuous, low-cost, weather-independent and almost real-time availability monitoring of water vapour.

In this chapter, several methods are introduced that serve to increase the spatial resolution of the moisture and temperature fields at the near-surface and in the boundary layer. Retrievals of column-integrated water vapour from GPS data are used to increase the resolution of the boundary layer moisture distribution. Most of the atmospheric water vapour is in the lower troposphere, and its variability is expected to be dominated by boundary layer variations, as we will demonstrate later. However, it is necessary to develop a new algorithm to relate the total-column water vapour information delivered by GPS to the boundary layer moisture distribution. For the first time to the author's knowledge, a new algorithm is introduced to combine the GPS IWV measurements with radiosonde moisture profiles to increase the spatial resolution of the water vapour field in the atmosphere. Furthermore, profiting from the high spatial resolution AWSs network at the near-surface measuring temperature, moisture and wind, high-resolution fields of these parameters are obtained. One of the primary goals of this thesis is to investigate whether an increase in resolution by AWS and/or GPS data leads to a significant improvement in the detection of the initiation of convection. Data from the CSIP IOP5, 29 June 2005, are used in this chapter to illustrate the analysis.

6.1 Interpolation scheme

To interpolate the available temperature, moisture and wind measurements from the surface and radiosonde stations onto a regular grid, the Inverse Square Distance Method (ISDM) (Shepard, 1968) is applied. Six radiosondes 25 to 50 km away from each other are available in the investigation area. To increase the resolution at the near-surface, measurements from AWSs and SYNOP stations are used. The minimum distance between two of these stations is in the order of 5 km. To increase the resolution of the radiosonde-interpolated water vapour profiles, GPS measurements are used. The minimum distance between the GPS stations is approximately

15-20 km. Therefore, for the interpolation of the radiosondes, an intermediate grid resolution of approximately 11 km is used.

The ISDM assigns to each cell the weighted average of the stations in the neighbourhood of the cell. The weighting function is the inverse of the squared distance that separates the station from the cell. This means that stations that are close to the cell being interpolated will have more weight than those located far from the cell. This algorithm is based on the idea that irregularly distributed data points are better correlated the smaller the spatial distance between them. The interpolated values on the regular grid are calculated according to,

$$\psi_{GP} = \frac{\Sigma_{i=1}^{n} \psi_i \frac{1}{d_i^2}}{\Sigma_{i=1}^{n} \frac{1}{d_i^2}} \tag{6.1}$$

where i indicates the specific stations, n is the total amount of stations, d_i being the distance between the station and the grid point, ψ_i is the measured and ψ_{GP} the interpolated value where the subscript GP stands for grid point. Many different interpolation methods (Thiessen, 1911; Cressman, 1959; Barnes, 1964; Gandin, 1965; Wahba and Wendelberger, 1980; Daley, 1991; Daly et al., 1994; Hutchinson, 1995; Frei and Schaer, 1998) have been developed within the last century. An overview of interpolation methods can be found in Grayson and Bloeschl (2000). In this study, different interpolation methods have been tested and different grid resolutions have been compared. No significant improvements are observed when using a higher resolution. On the contrary, the calculation time for the analysis increases as the number of grid points does. The comparison between the performance of interpolation methods points out that the efficiency of the ISDM is very similar to that of other more sophisticated analysis tools in spite of the higher effort (Dorninger et al., 2008). These findings are in agreement with earlier comparisons of interpolation methods (Skok and Vrhovec, 2006). The temperature, humidity and wind vertical data points from the six radiosondes are interpolated at fixed levels, Δz=25 m, and 2 h intervals, Δt=2 h.

6.2 Temperature and moisture fields at increased resolution

Different techniques are applied to increase the near-surface and PBL resolution using the synergy of different instruments, AWS, SYNOP, and GPS stations, in the area of investigation. Fig.6.1 shows the CSIP domain and the position of these instruments. Two sources of additional data are used to increase the spatial resolution of radiosonde measurements, i) an increase in resolution of the temperature, moisture and wind field at the near-surface using AWSs and SYNOP stations, and ii) an increase in resolution of the moisture field in the PBL making use of the IWV data delivered from GPS stations.

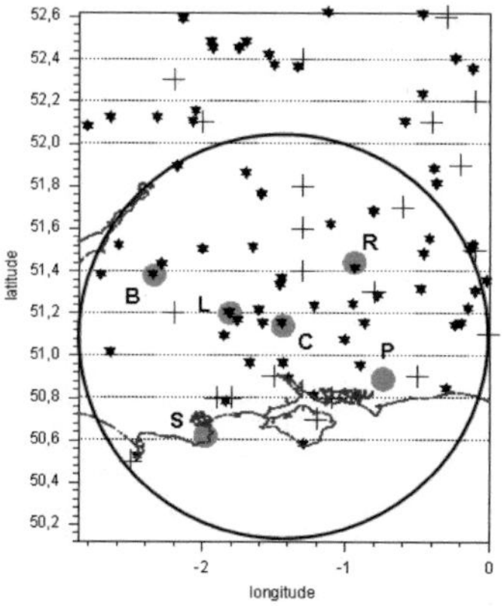

Figure 6.1: Locations of some of the instrumentation deployed in southern Britain during the CSIP campaign. The stars represent the positions of the AWS and SYNOP stations. The crosses indicate the positions of the GPS sites. The circles represent the positions of the six radiosonde stations Bath (B), Swanage (S), Larkhill (L), Chilbolton (C), Reading (R), and Preston Farm (P). The ring centered on Chilbolton indicates the 100 km range of the Chilbolton radar.

6.2.1 Increase in the resolution of the near-surface fields

When radiosondes are launched, they are initialized with temperature, humidity and wind observations measured at the near-surface. The radiosonde measurements closest to the surface are regarded as being representative of the situation at the near-surface. The resolution given by these measurements corresponds to the resolution given by the location of the radiosonde stations, which is in the order of 25 to 50 km. The interpolation method presented in section 1 is applied using the six radiosonde profiles in the area at 1000, 1200, and 1400 UTC. The measurements closest to the surface are selected and spatially plotted. As an example, the temperature and moisture distributions are shown in Fig.6.2.

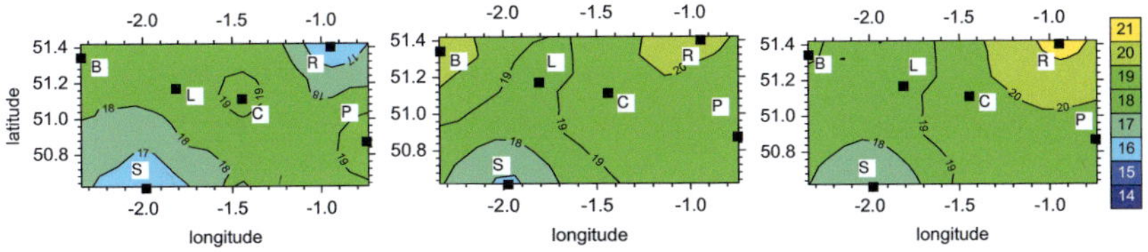

a)

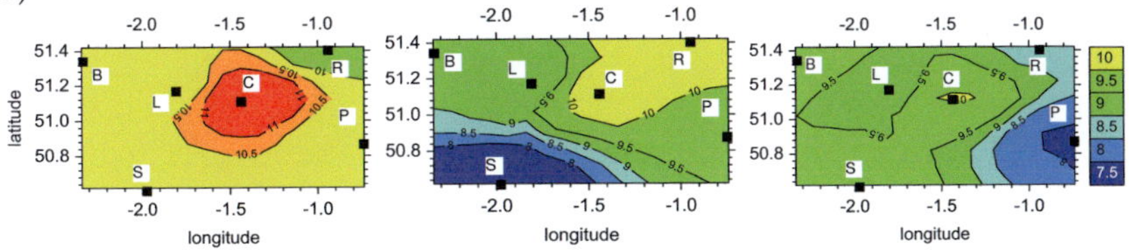

b)

Figure 6.2: Interpolated near-surface a) temperature in °C and b) mixing ratio in g kg^{-1}, at 1000 (left), 1200 (center) and 1400 UTC (right) for the CSIP domain on 29 June 2005 based on the six radiosondes in the area (black squares).

In addition to the AWSs and the SYNOP stations available from the Met Office, 25 AWS stations were deployed around Chilbolton (C) for the CSIP measurement campaign. Temperature at 2 m, humidity and wind measurements at 15 seconds temporal resolution are available. Temperature and moisture from the AWSs and SYNOP stations are interpolated into a regular grid using the ISDM interpolation scheme (Fig.6.3). For the interpolation, stations out of the boundaries are also used. This permits to extend the investigation area further to the north and west. The new investigation area extends up to 52.62° N and 2.65° W. For comparison between the low-resolution radiosonde interpolated fields (Fig.6.2) and the higher-resolution AWSs and SYNOP data interpolated fields (Fig.6.3), one should keep in mind that the investigation areas do not have the same extension.
The comparison shows that,

- The use of AWS and SYNOP stations allows the analysis of the near-surface conditions in areas where radiosondes are not available.
- The resolution of the AWS and SYNOP stations being much higher than the one obtained from radiosondes results in a much smaller-scale representation of the variability of the temperature, moisture and wind (shown later).
- In the areas where radiosonde measurements and AWS or SYNOP stations exist the measured values disagree. This may be a consequence of measure-

ments taken at different elevations or over different kinds of surfaces. However, the mesoscale gradients observed between the radiosonde stations are also reflected in the higher-resolution fields.

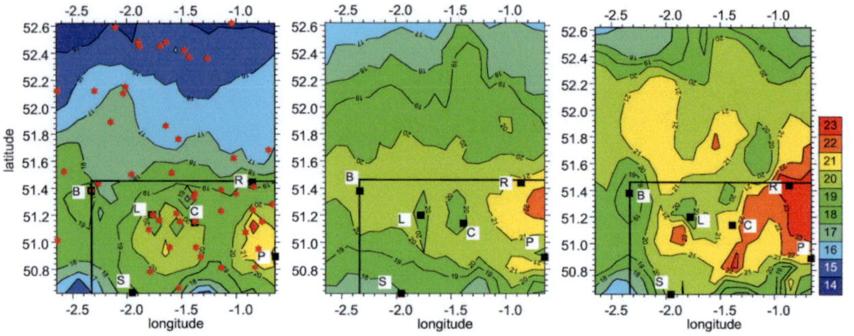

a)

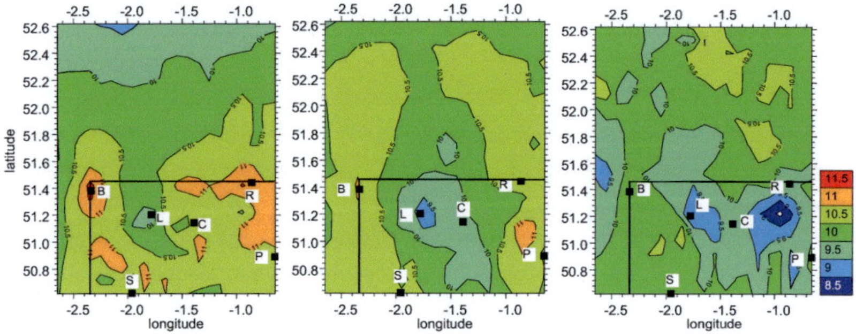

b)

Figure 6.3: Interpolated (a) temperature in °C and (b) mixing ratio in g kg^{-1}, at 2 m agl, at 1000 (left), 1200 (center) and 1400 UTC (right) from AWSs and SYNOP (red stars) measurements on 29 June 2005. Positions of the radiosonde stations (black squares). The area in Fig.6.2 is indicated within the black box.

6.2.2 Increase in the resolution of the water vapour fields

In order to obtain a spatial representation of the temperature and moisture distribution in the lower boundary layer, the mean of the lower 50 hPa is calculated from the radiosonde interpolated field (Fig.6.4). This quantity is important because convection indices are usually calculated by lifting a parcel with temperature and humidity characteristics averaged over the 50 hPa in the lower PBL.

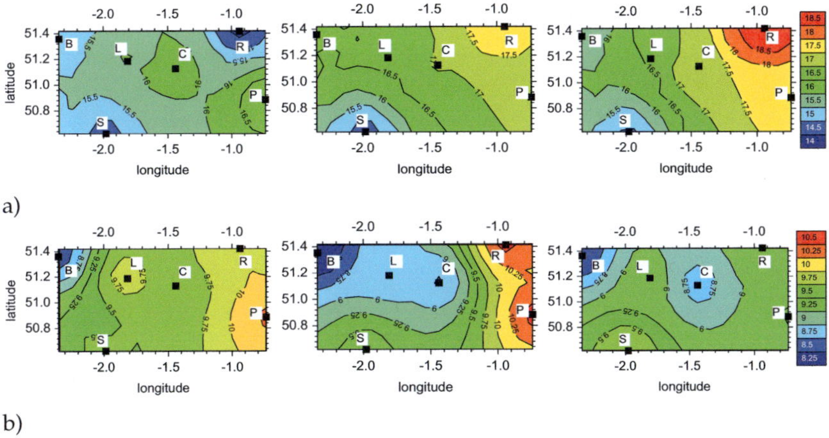

Figure 6.4: Radiosonde data interpolation of the 50 hPa mean of the lower PBL a) temperature in °C, and b) mixing ratio in g kg^{-1}, at 1000 (left), 1200 (centre) and 1400 UTC (right) on the 29 June 2005. Positions of the radiosonde stations (black squares).

In general, the temperature distribution in the lower boundary layer (Fig.6.4a) agrees with the near-surface observations (Fig.6.3a), showing increasing temperatures from west to east. However, the PBL moisture spatial distribution (Fig.6.4b) shows a different pattern than at the near-surface (Fig.6.3b). At the near-surface, the eastern area close to Bath is a region of high moisture with respect to the rest of the stations, whereas the 50 hPa PBL mean shows that the lower PBL at Bath is the driest in the area. The horizontal interpolation of the radiosonde data, despite the dense network deployed for the CSIP campaign, gives us a representation of the temperature and humidity horizontal variations in a scale of approximately 25 to 50 km. Smaller-scale variations, especially of the highly variable water vapour (Fig.5.3), are not represented. To obtain a more accurate representation of the PBL moisture distribution, the water vapour measurements from radiosondes and GPS measurements are combined. Some information about the GPS IWV measurements and their utility is given first.

GPS water vapour measurements

The GPS is an accurate and all-weather technique for the retrieval of IWV in ground-based networks (Bevis et al., 1992; Businger et al., 1996; Duan et al., 1996; Wickert and Gendt, 2006). Because the GPS signal is sensitive to the refractive index of the atmosphere, and this index is a function of pressure, temperature, and moisture, GPS can be used directly for sensing properties of the atmosphere. Small amounts of atmospheric water vapour significantly affect GPS signal propagation velocities (Rocken et al., 1993). Thus, GPS is especially suited for sensing tropospheric water vapour. The basic GPS atmospheric information results from the delay of the GPS signal that has traveled between a GPS satellite (at an altitude of 20200 km) and a ground-based receiver with respect to its propagation in the vacuum. The standard procedure for GPS data analysis assumes that the delay in any direction can be mapped from the delay at zenith, to which a horizontal gradient is added. Three sets of parameters are then estimated during the analysis: zenith tropospheric delays (ZTDs), gradients, and postfit residuals, which are the difference between the modeled atmosphere and the measurements (Bastin et al., 2007).

The ZTD has two components; the wet delay which is caused by atmospheric water vapour (ZWD), and the dry or hydrostatic delay (ZHD) caused by the remaining atmospheric constituents,

$$ZTD = ZHD + ZWD. \tag{6.2}$$

From Eq.6.2, the dry-atmosphere component (ZHD), depending only on the total pressure and the temperature, is removed and the reminder is converted into IWV. Following Yuan et al. (1993), the ZHD can be accurately estimated from the surface pressure P_0 and the variation of the gravity field $f(',H)$ with the latitude φ and the height H in kilometer (Davis et al., 1985; Saastamoinen, 1972),

$$ZHD = [(0.0022768 \pm 0.0000015) mhPa^{-1}] \frac{P_0}{f(',H)}, \tag{6.3}$$

with

$$f(',H) = 1 - 0.00265cos(2') - 0.000285H. \tag{6.4}$$

ZTD is thus mapped into IWV, using simply surface pressure and temperature and empirical formulas (Davis et al., 1985; Bevis et al., 1992; Emardson and Derks, 1999). ZWD is simply the difference between the ZTD and the ZHD. The Zenith IWV is nearly proportional to the ZWD, with a conversion factor Π, in kg m^{-3}, expressed after Bevis et al. (1992) as,

$$\Pi = \frac{10^6 m_w}{(k_2 - k_1 \frac{m_w}{m_d} + \frac{k_3}{T_m}) R^*}, \tag{6.5}$$

where T_m is the so-called "mean temperature", vertically integrated mean temperature within an atmospheric water vapour column represented by N levels, m_w is

the molar mass of water vapour, m_d is the molar mass of dry air, and R^* is the ideal gas constant. The relation between the ground temperature and the mean temperature T_m can be empirically determined from a data set of regional radiosoundings with an accuracy of 2% (Bevis et al., 1992).
Finally, we have,

$$IWV = \Pi\, ZWD. \tag{6.6}$$

The accuracy in GPS IWV has been assessed by a number of authors, using intercomparisons with radiosondes, microwave radiometers, sun photometers, lidars, and VLBI (Very Long Baseline Interferometry) (Bock et al., 2004). The agreement between these techniques is about 1-2 kg m^{-2}.

The Global Positioning System network of the United Kingdom

The UK Global Positioning System (GPS) data network is processed by the Met Office / University of Nottingham system based in Exeter and delivers retrievals of column-integrated water vapour at a fifteen-minute resolution. For the CSIP measurement campaign, additional five sites were installed and processed by GeoForschungsZentrum (GFZ), Potsdam. These sites were processed at thirty-minute resolution. The short temporal resolution of the GPS measurements is a significant advantage.

Combination of IWV measurements from radiosondes and GPS and spatio-temporal distribution of vertically integrated water vapour

For investigation of the spatial distribution and temporal evolution of the atmospheric water vapour in the investigation area, GPS IWV data is used. To increase the resolution of the GPS derived IWV field, IWV is calculated from the radiosondes in the area and both data sets combined.
This combination is made as follows,

1. Calculation of the IWV from the radiosondes.
2. Determination of a possible bias between the radiosondes and GPS measurements.
3. Combination and spatial interpolation of the IWV from radiosondes and GPS.

A detail description of these steps is given in the following,

1. Calculation of the IWV from radiosondes:
To calculate the column-integrated water vapour from radiosondes the following

integration is applied,

$$IWV = \int_0^z \rho_v(z)dz, \tag{6.7}$$

where z indicates the height of the column and ρ_v is the density of water vapour or absolute humidity in kg m^{-3}. For the calculation of the total-column water vapour, the radiosondes have been linearly extrapolated from the top measurement level up to the height where the pressure is approximately 100 hPa, where the water vapour content is assumed to be zero. ρ_v is computed as,

$$\rho_v = \frac{e}{R_v T} = \frac{e}{p}\epsilon\rho_d, \tag{6.8}$$

where ρ_d is the density of dry air, ρ_d=1.225 kg m^{-3} at sea level, and varies with altitude, pressure and temperature according to the ideal gas law, $\epsilon = R_d/R_v = 0.622$ is the ratio of gas constants for dry air to that for water vapour, e is the partial pressure of water vapour which is calculated at each level in the sounding from the dewpoint, temperature and pressure.
Once ρ_v values are calculated for each level in the sounding, IWV is calculated applying Eq.6.7. Precipitable water then is,

$$PW = \frac{IWV}{\rho_w}, \tag{6.9}$$

where ρ_w is the density of liquid water, ρ_w=1000 kg m^{-3}.
For evaluating numerically the integral in Eq.6.7, an elementary numerical integration method is used, the "Trapezium rule" whose error is approximately given by,

$$E \approx \frac{1}{12}nh^3 < f'' >= \frac{1}{12}(b-a)h^2 < f'' >, \tag{6.10}$$

where n is the number of vertical steps, h is the vertical step and $< f'' >$ represents an average value for the second derivative of the integral function, f, over the interval a to b. The estimated error using vertical steps of 25 m is in the order of 10^{-4} kg m^{-2}, which means that the error of this calculation is negligible in comparison to the errors associated with the radiosonde measurements.

2. Determination of the bias between radiosondes and GPS:
Radiosonde water vapour profiles are often considered "truth", but inconsistencies between radiosonde profiles recorded by the same type of instrument have been repeatedly noted (Turner et al., 2003). Thus, the development of correction algorithms has been required (Lucas and Zipser, 2000; Cady-Pereira et al., 2007). The Vaisala RS90 and RS92 radiosondes are some of the most commonly deployed sonde profilers. Their accuracy can be greatly improved by scaling the total column PW by the PW retrieved from the IWV delivered by the GPS. Thus, to combine

GPS derived IWV and radiosonde calculated IWV data, it is necessary to estimate a possible bias. The different locations of GPS and radiosondes excluded direct comparison of the GPS-derived IWV data with the IWV data calculated from the radiosondes. Nevertheless, two alternative methods have been used to be able to estimate the possible bias between both instruments and compare it with former studies.

i) In the first method, we make use of the two radiosonde and GPS stations that are closer to each other; IWV from Reading radiosonde station (51.44°N, 0.94°W) is compared with IWV from FARN GPS station (51.3°N, 0.8°W) and IWV from Bath radiosonde station (51.38°N, 2.34°W) with WARM GPS station (51.2°N, 2.2°W). A distance of approximately 18 km separates Reading radiosonde station and FARN GPS station, and between Bath and WARM the distance is nearly 22 km. Data from IOP2 (radiosonde profiles at Bath are not available), IOP5 and IOP8, and measurements at 1000, 1200 and 1400 UTC are used. For comparison, different possibilities were tested, some of which are described as follows: In Fig.6.5, the IWV from Reading and Bath calculated from the radiosondes are plotted against IWV measurements from GPS stations at FARN and WARM, respectively. The IWV values calculated from radiosondes and from GPS differ, e.g for 13 July 2005, IOP8, the IWV from the radiosonde is ~ 34 kg m^{-2} whereas the GPS IWV measurement is ~ 31 kg m^{-2}. This difference, which is greater than the error in the GPS measurements, might suggest that a bias exists between both instruments.

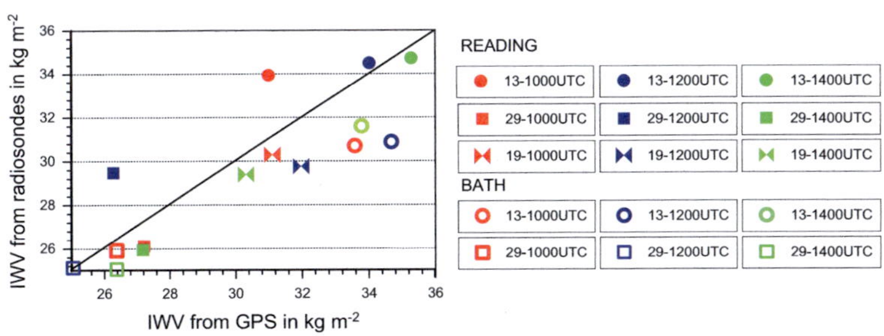

Figure 6.5: IWV measurements from radiosondes vs IWV from GPS for IOP2, 19 June 2005, (inverted triangles), IOP5, 29 June 2005, (squares) and IOP8, 13 July 2005, (circles) at 1000 (red), 1200 (blue) and 1400 UTC (green). Measurements from Reading are plotted with full marks while those from Bath are indicated by open marks.

COMPARISON	IWV_{GPS} mean in kg m^{-2}	IWV_{RAD} mean in kg m^{-2}	bias in %
all IOPs; all times	30.3	29.4	2.7 *Dry*
Bath; IOP5, IOP8; all times	29.9	27.9	6.6 *Dry*
Reading; IOP5, IOP8; all times	30.2	30.7	1.9 *Wet*
Reading; all IOPs; all times	30.5	30.4	0.2 *Dry*
IOP8; all stations; all times	33.7	32.7	3.0 *Dry*
IOP5; all stations; all times	26.4	26.0	1.3 *Dry*
IOP2; Bath; all times	31.1	29.8	4.3 *Dry*
1000UTC; all stations; all IOPs	29.8	29.4	1.6 *Dry*
1200UTC; all stations; all IOPs	30.3	29.9	1.4 *Dry*
1400UTC; all stations; all IOPs	30.6	29.0	5.1 *Dry*

Table 6.1: IWV comparisons between calculations from radiosondes and measurements from the closest GPS stations. Dry bias indicates that the IWV calculations from radiosondes show less water vapour content than the GPS measurements.

Some of the biases obtained from the different comparisons are summarized in Tab.6.1. A large variability is observed, from 1.9 % wet bias to a 6.6 % dry bias, therefore, no simple correction can be obtained from these comparisons. A reason for the large variability observed may be that the distance of approximately 20 km between radiosondes and GPS stations is so large that air masses of different humidity are present. The GPS measurements are based on samples of the atmosphere forming a cone, where the area on the ground is less than 1 m^2 and increasing with height at a rate determined by the specified elevation cutoff angle. This result indicates that large water vapour variability is present within 20 km.

ii) As a second attempt, the IWV calculated from radiosondes is spatially interpolated using ISDM, and the GPS-derived IWV data is compared directly with the radiosonde-interpolated IWV data at the same location of the GPS station (Fig.6.6). The stations, IOPs and times correspond to those used in the first method.

From this comparison, an average radiosonde dry bias of approximately 6.5% is obtained in agreement with other intercomparisons between total column precipitable water vapour and radiosonde humidity sensors where the comparisons yielded an average 5% dry bias (Guichard et al., 2000; Cady-Pereira et al., 2007). This daytime dry bias has been attributed to solar heating of the sensor, and is more evident in the RS90 and RS92 profiles due to the absence of a cap on the sensors, which was used on the RS80s to shield the sensor from precipitation and solar radiation. Former experiments found that the daytime dry bias increases with altitude and can reach 50% near the tropopause (Cady-Pereira et al., 2007). To correct the dry bias from the radiosonde data, the radiosonde moisture profile, up to 100 hPa, is corrected by a factor of 5%. To quantify the effect that moisture correction exerts on convection indices, these are calculated and compared for the uncorrected and corrected profiles. Results for the two radiosonde stations used for the comparison, Reading and Bath, are shown in Table.(6.2). The dry bias in the radiosonde humidity profiles led to the underestimation of CAPE and LI and an overestimation of CIN and CAP, in agreement with former studies (Guichard et al., 2000; Cady-Pereira et al., 2007). Correcting the measured humidity profiles yielded more accurate values of atmospheric instability and inhibition and a more realistic behaviour.

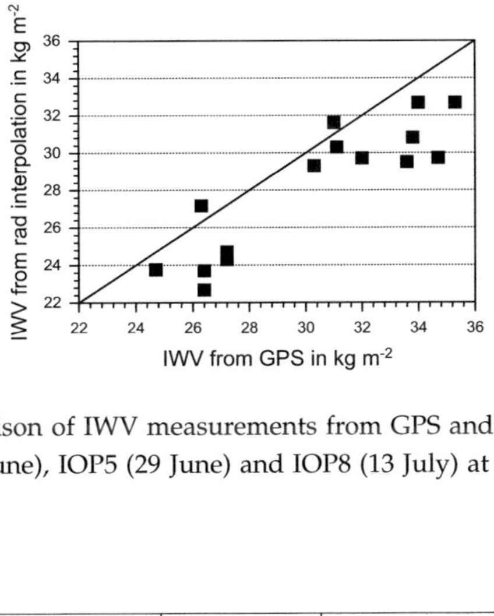

Figure 6.6: Comparison of IWV measurements from GPS and radiosonde calculations for IOP2 (19 June), IOP5 (29 June) and IOP8 (13 July) at 1000, 1200 and 1400 UTC.

1200 UTC	R uncorrected	R corrected	B uncorrected	B corrected
CAPE	554	763	153	194
CIN	-2	-1	-9	-8
LI	-2.3	-3.1	1.1	0.6
CAP	0.4	0.4	1.3	1.0

Table 6.2: Effect of water vapour profile correction on convection indices. CAPE in J kg^{-1}, CIN in J kg^{-1}, LI in °C and CAP in °C calculations for Reading (R) and Bath (B) radiosonde stations at 1200 UTC on 29 June 2005 are presented. The convection indices are calculated lifting a parcel with temperature and humidity characteristics averaged over the lowest 50 hPa of the PBL.

3. Spatio-temporal distribution of IWV:
Finally, once the instrumental bias between the radiosondes and the GPS IWV measurements is corrected, the data can be combined. From this combination, high-resolution IWV fields are obtained by ISDM interpolation of the data. In Fig.6.7, the spatio-temporal evolution of IWV as a combination of GPS measurements and radiosonde calculations is presented for the area of investigation in the period from 1000 to 1500 UTC. The southwest-to-northeast gradient shown by the GPS retrievals in the CSIP domain is consistent with the radiosonde observations of a wetter boundary layer at Reading compared with the other radiosonde sites (Fig.6.4b). This gradient in column-integrated water vapour was present from 0830 UTC (not shown) to at least 1400 UTC.

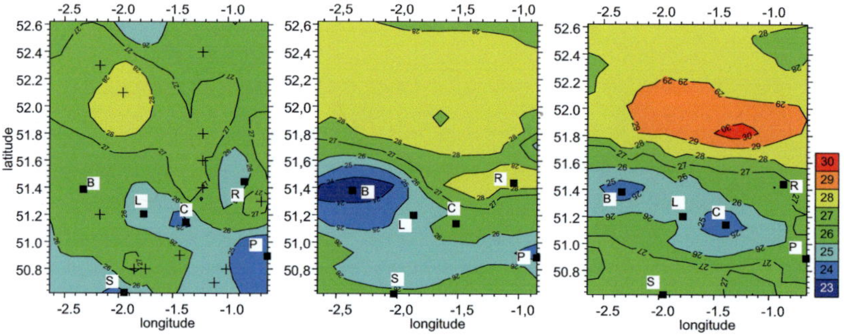

Figure 6.7: Spatio-temporal evolution of IWV in kg m^{-2} on 29 June 2005 at 1000 UTC (left), 1200 UTC (center), and 1400 UTC (right) with positions of the radiosonde stations (black squares) and positions of the GPS stations (stars).

Vertically resolved water vapour distribution in the atmosphere

Since IWV is an integrated value over the total column, there is no implicit information about the vertical distribution of moisture in the different layers of the atmosphere. To estimate it, height-resolved data from radiosondes is used. The water vapour content in the six radiosonde sites is integrated following Eq.6.7 for different layers in the atmosphere to quantify the water vapour stratification and the temporal variations in the different layers. With this purpose, water vapour content is calculated for the total column and different atmospheric layers (Tab.6.3). The radiosonde stations are not at the same altitudes due to the topography of the area. To take into account these differences, a normalized IWV is computed for the 900hPa-surface layer and the 950hPa-surface layer by dividing the IWV by the difference between the pressure at the surface level and the pressure at the top of the layer.

STATIONS 1000 UTC	IWVtotal column	IWV 300 hPa to 500 hPa	IWV 500 hPa 5 to 900 hPa	IWV 900 hPa to surf	%	IWV 950 hPa to surf	%
Reading	26.1	0.7	16.2	9.1	35.1	5.3	20.3
Chilbolton	26.5	1.1	16.1	8.7	32.7	4.5	17.0
Larkhill	27.0	1.1	16.8	8.9	33.3	4.3	15.9
Bath	25.9	1.1	16.2	7.4	28.6	3.2	12.4
Swanage	25.7	0.9	15.4	9.2	36.0	5.3	20.7
Preston	25.2	0.9	14.9	9.3	37.0	5.3	21.2

STATIONS 1200 UTC	IWVtotal column	IWV 300 hPa to 500 hPa	IWV 500 hPa 5 to 900 hPa	IWV 900 hPa to surf	%	IWV 950 hPa to surf	%
Reading	29.4	1.0	17.9	10.4	35.3	5.7	19.5
Chilbolton	26.9	1.2	16.8	8.1	30.4	4.1	15.4
Larkhill	26.5	1.3	17.1	8.1	30.3	3.8	14.5
Bath	25.1	1.3	16.1	7.1	28.6	3.1	12.6
Swanage	27.2	1.1	16.5	9.5	35.0	5.3	19.6
Preston	25.4	1.1	14.6	9.7	38.2	5.7	22.5

STATIONS 1400 UTC	IWVtotal column	IWV 300 hPa to 500 hPa	IWV 500 hPa 5 to 900 hPa	IWV 900 hPa to surf	%	IWV 950 hPa to surf	%
Reading	25.9	0.9	15.6	9.3	36.0	4.7	18.3
Chilbolton	23.6	0.1	14.5	8.0	34.0	4.0	17.2
Larkhill	25.8	0.9	16.1	8.1	31.5	3.9	15.4
Bath	23.6	1.0	15.3	7.1	30.3	3.2	13.5
Swanage	27.4	1.1	16.7	9.6	35.1	5.4	19.6
Preston	26.6	1.0	16.4	9.2	34.6	4.8	18.1

Table 6.3: IWV calculations from radiosondes at 1000, 1200 and 1400 UTC on 29 June 2005.

From these results important conclusions are derived,

- The IWV variability is dominated by boundary layer variations, as we expected. Between 28 % to 40 % of the water vapour is below 900 hPa, and between 12.5 % to 22.5 % is below 950 hPa.

- Between 500 and 300 hPa water vapour content varies between approximately 0.7 to 1.3 kg m^{-2} for this time period.

- Reading is the only station where an increase in IWV is observed in all layers from 1000 to 1200 UTC, especially below 900 hPa.

Methods to relate IWV and water vapour in the boundary layer

In this section, IWV from GPS is used to increase the resolution of the radiosonde-interpolated water vapour profiles. For this purpose, new algorithms are introduced and discussed.

The steps followed before aplying the proposed algorithms are enumerated in the following,

1. Radiosonde-interpolated profiles are obtained as explained in section 1 of this chapter.

2. The IWV is calculated for each radiosonde-interpolated profile following Eq.6.7.

3. A bias between GPS-derived IWV (IWV^{GPS}) and radiosonde-calculated IWV (IWV^{RAD}) is determined and corrected.

4. IWV^{GPS} and IWV^{RAD} are combined (IWV$^{RAD+GPS}$) and interpolated into a regular grid, resulting in a higher-resolution field of IWV (Fig.6.8).

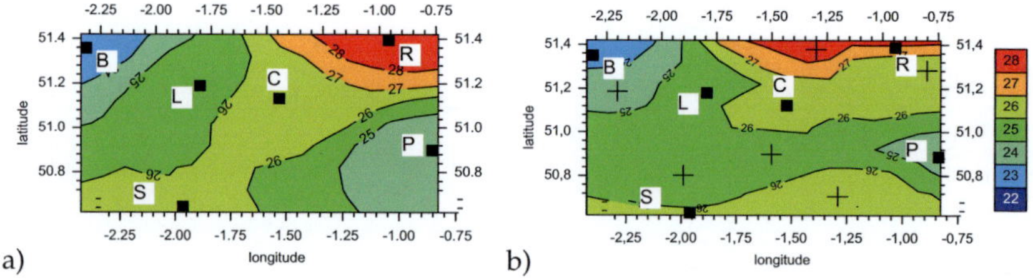

Figure 6.8: a) IWV interpolated field from radiosondes (low resolution), and b) IWV interpolated field from a combination of radiosondes and GPS measurements (higher resolution) on 29 June 2005 at 1200 UTC. Black squares indicate the positions of the six radiosonde stations and stars indicate the positions of the GPS stations.

Station Station	IWV_{RAD} in kg m^{-2}	r_{RAD}^{50hPa} in g kg^{-1}	r_{RAD}^{50hPa} / IWV_{RAD} in g kg^{-2} m^2
Bath	25.0	8.4	0.335
Chilbolton	26.9	8.9	0.333
Larkhill	26.5	8.9	0.338
Preston F.	25.5	10.5	0.412
Reading	29.4	10.6	0.360
Swanage	27.2	9.7	0.356

Table 6.4: Ratio between the mixing ratio mean of the lower 50 hPa in the boundary layer and the column-integrated water vapour from radiosondes on 29 June 2005 at 1200 UTC.

Since IVW and mixing ratio are both a measure of the water vapour content in the atmosphere, a first hypothesis is to assume that the ratio between them is constant at a certain layer. In particular the mean mixing ratio of the 50hPa of the lower PBL, r_{50hPa}, is used. Initially, a ratio is obtained between the mixing ratio 50hPa mean of the lower boundary layer, r_{50hPa}, and the column-integrated water vapour, from radiosondes, IWV_{RAD} (Tab.6.4). The spatial dependency of the water vapour distribution is taken into account to obtain a high-resolution interpolated field which reproduces the radiosonde measurements at the locations of these and which does not change drastically the whole distribution if radiosondes are removed or additional ones are used (Fig.6.9).

In particular, for the 50 hPa mixing ratio mean,

$$r_{50hPa}^{GPS+RAD}(x,y) = \frac{r_{50hPa}^{RAD}(x,y)}{IWV^{RAD}(x,y)} IWV^{GPS+RAD}(x,y). \tag{6.11}$$

Applying Eq.6.11, a new mixing ratio field in g kg^{-1} is obtained (Fig.6.9). The relation between the IWV and water vapour mixing ratio can be calculated for each layer in the atmosphere. However, we rely on an intuitive assumption about the relation between IWV and the water vapour distribution, in this case, the mean mixing ratio in the different atmospheric layers. Therefore, a more reliably method to relate IWV and water vapour mixing ratio in the different atmospheric layers is needed. Such a method has been called *adjustment method*. It is based on the assumption that the combined high resolution $IWV^{RAD+GPS}$ field can be used to adjust the interpolated field with radiosonde humidity profiles, because it is claimed that the vertically integrated water vapour from the radiosonde fields must agree with the combined $IWV^{RAD+GPS}$ values at each grid site. This agreement is by

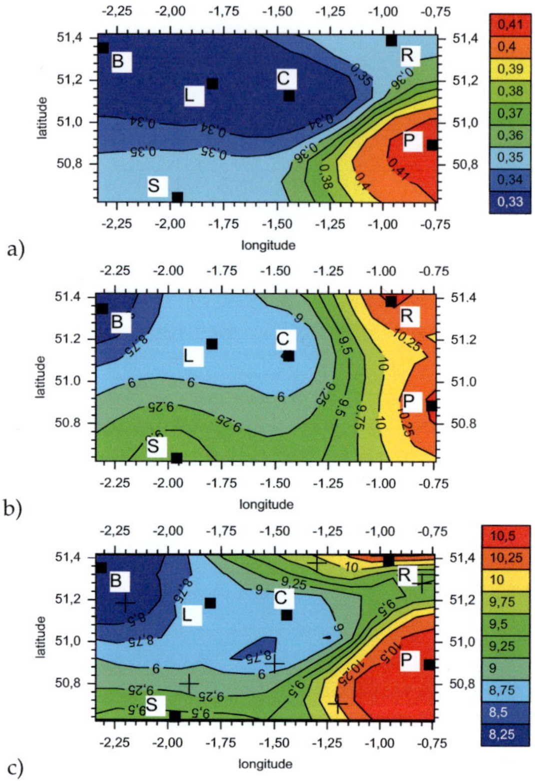

Figure 6.9: a) Spatial distribution of the ratio $r^{RAD}_{50hPa}/IWV^{RAD}$ field, and b) water vapour mixing ratio mean of the lowest 50 hPa of the PBL in g kg^{-1} from interpolation of radiosondes, and c) water vapour mixing ratio mean of the lowest 50 hPa of the PBL in g kg^{-1} from combination of radiosonde and GPS data. The positions of the radiosondes are indicated by black squares and the positions of the GPS stations are indicated by stars.

definition valid for the radiosonde sites but may not hold for the grid points in between. Therefore, for these grid points a difference (DIF) may exist.
This method is applied as follows,

1. For each grid point, a comparison is made between IWV^{RAD} the low-resolution field (from interpolated radiosonde profiles) and $IWV^{RAD+GPS}$ the high-resolution field (from interpolation of IWV from radiosondes and GPS) (Fig.6.8).

2. From this comparison, an IWV difference is obtained for each grid point between the high-resolution and the low-resolution fields,

$$IWV^{RAD+GPS}(x,y) = IWV^{RAD}(x,y) + DIF(x,y). \tag{6.12}$$

   This difference is normally different from zero, except at the radiosounding sites.

3. The integral expression for the calculation of the IWV is given by,

$$IWV^{RAD}(x,y) = \int_0^z \rho_v^{RAD}(x,y,z)dz, \tag{6.13}$$

$$IWV^{RAD+GPS}(x,y) = \int_0^z \rho_v^{RAD+GPS}(x,y,z)dz. \tag{6.14}$$

   Formally, the "DIF(x,y)" term can be written as,

$$DIF(x,y) = \int_0^z f(x,y,z)dz = \int_0^z \underbrace{\left(\frac{f(x,y,z)}{\rho_v^{RAD}(x,y,z)}\right)}_{AF(x,y,z)} \rho_v^{RAD}(x,y,z)dz, \tag{6.15}$$

   where $f(x,y,z)$ is an unknown function which, integrated, gives the measured DIF(x,y), and AF(x,y,z) is the so-called *Adjustment Factor*.

   By introducing this integral equation, we are able to relate two measurable quantities, DIF(x,y) and $\rho_v^{RAD}(x,y,z)$, at the cost of introducing the unknown AF(x,y,z). However, if we assume now that the dependence on z cancels in the ratio f / ρ_v^{RAD}, then AF(x,y,z) is independent from z, thus AF(x,y,z)=AF(x,y). This assumption implies that the dependence of f(x,y,z) in z is exactly the same of $\rho_v^{RAD}(x,y,z)$ and one may extract the *Adjustment Factor* from the integral. Therefore, Eq.6.12 can be written as,

$$\int_0^z \rho_v^{RAD+GPS}(x,y,z)dz = \int_0^z \rho_v^{RAD}(x,y,z)dz + AF(x,y)\int_0^z \rho_v^{RAD}(x,y,z)dz, \tag{6.16}$$

   and integrating,

$$IWV^{RAD+GPS}(x,y) = IWV^{RAD}(x,y) + AF(x,y)IWV^{RAD}(x,y). \tag{6.17}$$

From Eq.6.12 and Eq.6.17,

$$AF(x,y) = \frac{DIF(x,y)}{IWV^{RAD}(x,y)} = \frac{IWV^{RAD+GPS}(x,y) - IWV^{RAD}(x,y)}{IWV^{RAD}(x,y)}, \quad (6.18)$$

where the term on the right is known and AF can be estimated.

4. Once the AF is calculated from Eq.6.18, the higher resolution water vapour profiles are obtained. For each vertical data point in the moisture profile,

$$\rho_v^{new}(x,y,z) = \rho_v^{RAD+GPS}(x,y,z) = \rho_v^{RAD}(x,y,z) + \underbrace{AF(x,y)\rho_v^{RAD}(x,y,z)}_{adjustment}, \quad (6.19)$$

where the second term on the right side of the equation is an adjustment for each data point in the moisture profile. It is important to notice that the adjustment applied at each vertical level depends on the amount of water vapour present for each level (Fig.6.10).

The difference between this method and the former one lies in the fact that in this case we have a systematic way to incorporate the GPS data into our profile and if experimentally we can obtain some information of the z dependence of the ratio that defines the AF(x,y,z) factor, this could be directly incorporated. Using the second term on the right of Eq.6.18, Eq.6.19 can be written in the form,

$$\rho_v^{RAD+GPS}(x,y,z) = \frac{\rho_v^{RAD}(x,y,z)}{IWV^{RAD}(x,z,y)} IWV^{GPS+RAD}(x,y), \quad (6.20)$$

which is exactly the same relation that we found for the mixing ratio in Eq.6.11 by using an intuitive argument. However, to recover Eq.6.11, we have to further assume that e is much smaller that p in the whole profile. Applying a Taylor expansion, one obtains,

$$r_{50hPa} = \epsilon \frac{e}{(p-e)} \approx \epsilon \frac{e}{p} = \frac{\rho_v}{\rho_d}, \quad (6.21)$$

so that

$$\begin{aligned} r_{50hPa}^{new} &= \frac{\rho_v^{new}}{\rho_d} = \frac{1}{\rho_d} \frac{\rho_v^{RAD}(x,y,z)}{IWV^{RAD}(x,y,z)} IWV^{GPS+RAD}(x,y) \\ &= \frac{r_{50hPa}^{RAD}(x,y)}{IWV^{RAD}(x,y)} IWV^{GPS+RAD}(x,y). \end{aligned} \quad (6.22)$$

5. Once the *Adjustment Factor* has been applied to the radiosonde interpolated field, the water vapour mixing ratio distribution can be estimated for all atmospheric levels.

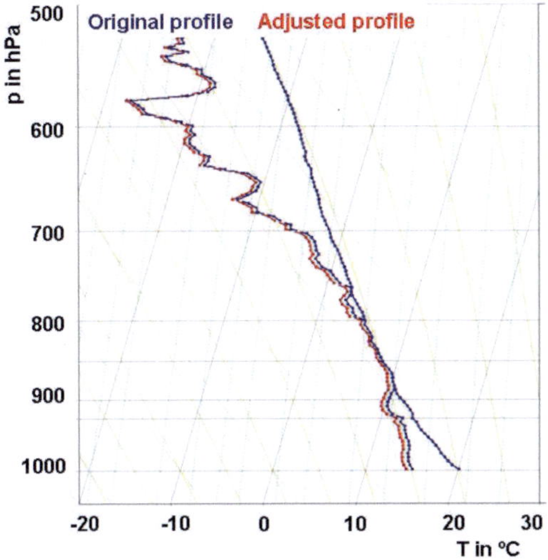

Figure 6.10: Skew T-log p plot showing temperature (right) and moisture profile (left) plus the adjusted moisture profile (red).

Applying the adjustment method, a higher-resolution field of water vapour profiles is obtained which is important in order to analyze water vapour horizontal and vertical distribution in the atmosphere and its relation with the initiation of convection. One important consequence of moisture adjustment is its effect on the calculation of the spatial distribution of convection indices. As discussed before, for the calculation of convection indices different methods can be used. Independently of the method used, good moisture profiles are of advantage. When applying virtual temperature correction (Doswell III and Rasmussen, 1994), the adjustment in the moisture profiles for the radiosonde-interpolated profiles is even more important since it is the comparison between the virtually corrected environmental temperature profile and the virtually corrected lifted profile which is used for the estimation of most of the convection indices. Therefore, a moisture adjustment at all radiosonde-interpolated vertical levels implies also an adjustment in the virtually corrected radiosonde-interpolated temperature profile.

Chapter 7

High Spatial Resolution Fields of Convection-Related Indices and their impact on the detection of the initiation of convection

The algorithms introduced in the last chapter are applied to achieve high-resolution fields of temperature and water vapour on a regular grid. An increase in resolution of the thermodynamic fields also results in an increase in resolution in the spatial representation of the convection indices and convection inhibition parameters. One goal in this chapter is to ascertain whether this increase in resolution suffices to adequately represent the location of the initiation of deep convection. Furthermore, based on the idea that a set of conditions is required for each weather phenomenon to occur (Doswell et al., 1996), it is the main goal in this chapter to find an optimal combination of convection-related indices to describe convection initiation. The working hypothesis is that the likelihood of deep convection is high for regions with high conditional instability, low atmospheric inhibition, absence of mid-troposopheric lids, and the presence of a triggering mechanism. CSIP and COPS data are used to verify this hypothesis. A detailed description of the convection indices used in this chapter and their computation was presented in Chapter 2. To illustrate the analysis, data from the CSIP IOP5, 29 June 2005, is used. Initially, the synoptic situation and convection activity on that day are discussed.

7.1 IOP5: synoptic situation and convection activity

The synoptic situation on 29 June 2005 is dominated by a mature cyclone present over the British Isles (Fig.7.1). The CSIP area at 1200 UTC is south of the occluded front associated with this cyclone. In southern UK, south-westerly winds in the PBL and southerly winds in the mid-troposphere are observed from the early morning onwards. Weak subsidence is observed in the CSIP area (Fig.7.1) so that synoptically forced deep convection could be excluded and initiation of convection must be dominated by boundary layer processes. The KO-index of about -4 K in the investigation area indicates a potentially unstable atmosphere (Fig.7.1). Showers in the CSIP area are observed by 1200 UTC (Fig.7.2). They became widespread slightly later than 1200 UTC within the CSIP area, and become intense enough to form a band of heavy precipitation extending east-west across Britain by around 1500 UTC (Fig.7.3). Because of this heavy rain, some flash flooding occurred at Oxfordshire to the north-east of the CSIP domain. At 1200 UTC, a cumulonimbus appears approximately 20 km north-east of Chilbolton (Fig.7.2). By 1300 UTC, there were two significant showers which have grown in size and have been advected northwards, intensifying when passing Reading (Fig.7.2). At this time, the satellite image shows the beginning of a line of cloud cells to the north of Chilbolton. By 1500 UTC, a line of well-organized storms is formed (Fig.7.3). Although only a few of the storms produced heavy precipitation in the CSIP area, the line of storms, approximately 100 km north-east of Chilbolton developed from showers that were initiated near this station. The initiation of convection occurred in the north-eastern region of the CSIP domain.

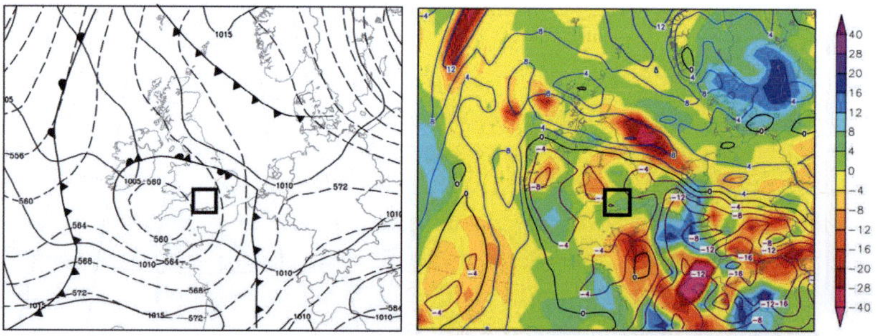

Figure 7.1: (a) The Global Forecast System (GFS) model analysis of sea level pressure (solid lines) and 500 hPa level in gpdm (dashed lines). (b) 500 hPa vertical motion in hPa h^{-1} (colour code) and KO-index in K (isolines) at 1200 UTC on 29 June, 2005. The investigation area is indicated by a square (all GFS charts courtesy of www.wetter3.de).

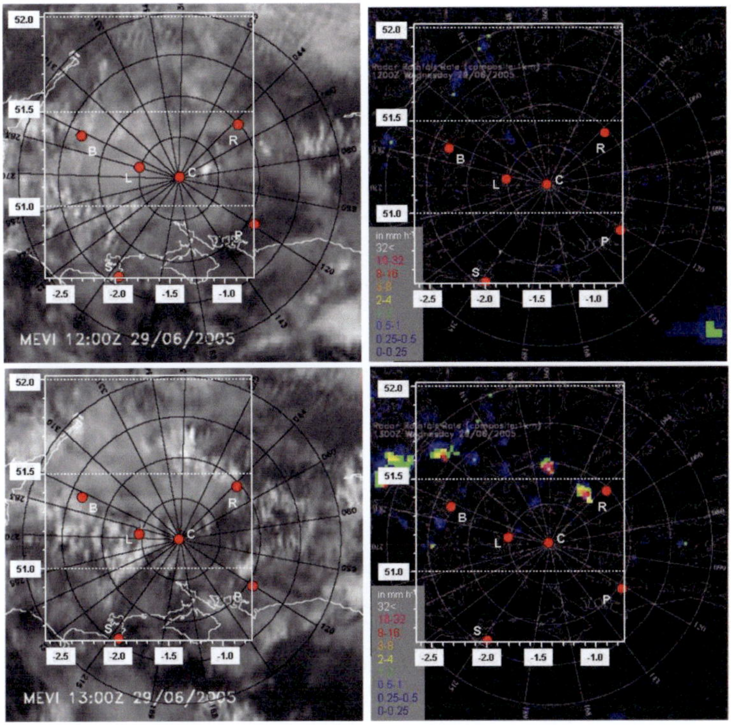

Figure 7.2: High-resolution visible imagery at 1200 UTC (top-left) and 1300 UTC (bottom-left) from Meteosat Secondary Generation (MSG) and radar rain rates in mm h^{-1} at 1200 UTC (top-right) and 1300 UTC (bottom-right) on 29 June, 2005. Range rings centered on the Chilbolton radar are shown at 25 km intervals. Positions of the radiosonde sites Bath (B), Swanage (S), Larkhill (L), Chilbolton (C), Reading (R), and Preston Farm (P) are indicated by red circles.

7.2 Evolution of convection-related indices

The algorithms introduced in the last chapter are applied to increase the resolution of temperature and water vapour fields and interpolate them on an 11 km grid. The convective indices are calculated using two different parcel definitions (i) using a surface-based (SB) parcel based on temperature and moisture from AWS and SYNOP and, (ii) using a parcel with 50 hPa mixed-layer (ML) characteristics. The analysis allows the investigation of whether the parcels are - with respect to the observation of deep convection - better defined in terms of the near-surface or averaged over the lowest 50 hPa of the PBL. As discussed in section 2.5, the absolute values of the convection indices generally differ when using different levels for the parcel to lift. The main interest lies in the spatial distribution of the convection indices. Figs.6.3 and 6.7 show that the surface and the GPS stations extend beyond the area covered by the radiosonde network. Nevertheless, temperature and moisture profile shapes are needed to apply the adjustment method (see section 6.6.2). Therefore, it is assumed that the temperature and humidity profile shapes of Bath and Reading are characteristic for the atmosphere further to the north for about 100 km and to the west for about 30 km. The humidity profiles are then adjusted corresponding to the available GPS data. This helps to interpret, with respect to the assumption made above, the evolution of deep convection north of the CSIP domain. In total, the investigation area extends from 2.65° W to 0.75° W and from 50.62° N to 52.62° N. Data at 1000, 1200 and 1400 UTC were analyzed, corresponding with the temporal availability of radiosondes during the period with convective development.

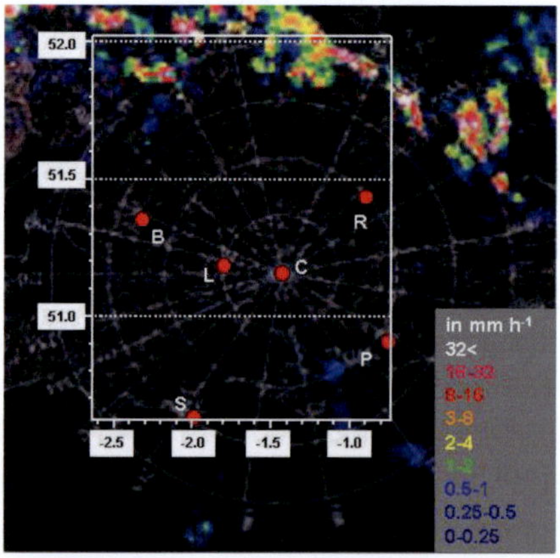

Figure 7.3: Radar rain rate at 1500 UTC on 29 June 2005. Range rings centered on the Chilbolton radar are shown at 25 km intervals. Positions of the radiosonde sites are indicated by red circles.

7.2.1 Distribution of CAPE

The spatio-temporal evolution of CAPE calculated with the surface-based and 50 hPa mixed-layer parcels are displayed in Fig.7.4. In addition to the CAPE values, tracks of the showers initiating in the area are also plotted. These were obtained by

Marsham et al. (2007b) using the rain-radar network and Meteosat satellite imagery to track the observed precipitation and the corresponding cumulus convection. It allowed the determination of the locations of convective initiation start of precipitation. The starting point of this analysis was to determine the trajectories of all precipitation echoes forming within the CSIP area that could be tracked for at least two frames of the 15 min data from the radar network. Those echoes that formed on the outer edge of pre-existing showers, which are likely to have been due to secondary initiation, were excluded from the analysis. Before 1130 UTC, no intiation is observed in the area. At 1200 UTC, isolated convective cells formed close to the Bath radiosonde station, and a cumulonimbus appears approximately 20 km north-east of Chilbolton. At 1400 UTC, new cells did initiate close to the pre-existing cells and hence were probably influenced by the convective outflow from the existing cells. Marsham et al. (2007a) discussed how cirrus patches left over from previous thunderstorms reduced sensible and latent heat fluxes in the CSIP area during CSIP IOP 5. They showed that in all but one of the cases, showers were observed initiating at the rear edges of the gaps, at the leading edge of the anvil, or in clear skies, pointing out that shading from cirrus anvils had a significant effect on convective initiation. We assume that these processes are reflected in the convection indices.

By 1000 UTC, low CAPE values were observed in most of the area except near Preston Farm, where CAPE $\sim$ 700 J kg^{-1}, when using the SB parcel (Fig.7.4a), and in the north of Bath at about 52° N, CAPE $\sim$ 500 J kg^{-1}, when using the ML parcel (Fig.7.4b). By 1200 UTC, an increase in CAPE is observed in most of the CSIP domain especially in the northern areas. In the southern region, around Bath, Larkhill and Swanage, the increase in CAPE is much lower. The most significant difference between the CAPE computed using the SB parcel and the 50 hPa ML parcel is found in the area of Bath where in the first case CAPE values of 1800 J kg^{-1} are present whereas when computing CAPE using the 50 hPa ML parcel it is below 300 J kg^{-1}. The high CAPE values observed in the northwest of the investigation area at about 52° N correspond to an area where precipitating cells are present (Fig.7.2), which were initiated west of the investigation area. By 1400 UTC, the CAPE values, calculated by both methods, remained low or even decreased south of 51.4° N but higher CAPE values were still observed in the northern areas. In particular, the use of the 50 hPa ML parcel choice clearly indicates a strong increase in CAPE in an east-west oriented band at about 52° N coinciding with the area where a well-organized line of thunderstorms with showers is observed at 1500 UTC (Fig. 7.3). In this area, CAPE has increased from 600-700 J kg^{-1} at 1200 UTC to values of about 1100 J kg^{-1} at 1400 UTC. This indicates the presence of strong conditional instability. The spatial-temporal evolution of CAPE can be mainly traced back to the evolution of temperature and moisture. Near to the surface, warming takes place in nearly all parts of the CSIP domain (Fig.6.3). Around noon, the highest values are observed along a line from Bath to Reading. Simultaneously, the moisture shows an increase in the northern parts while it decreases in most of the southern parts. At the near-surface, at 1200 UTC two areas of high moisture can be distinguished, one close to Preston Farm and Reading and one close to Bath. In the center of this region, around Chilbolton and Larkhill, a dry area is found. This change in moisture contributes to a shift of the maximum CAPE values towards the north (Fig.7.4). The overall development of temperature and moisture in the boundary

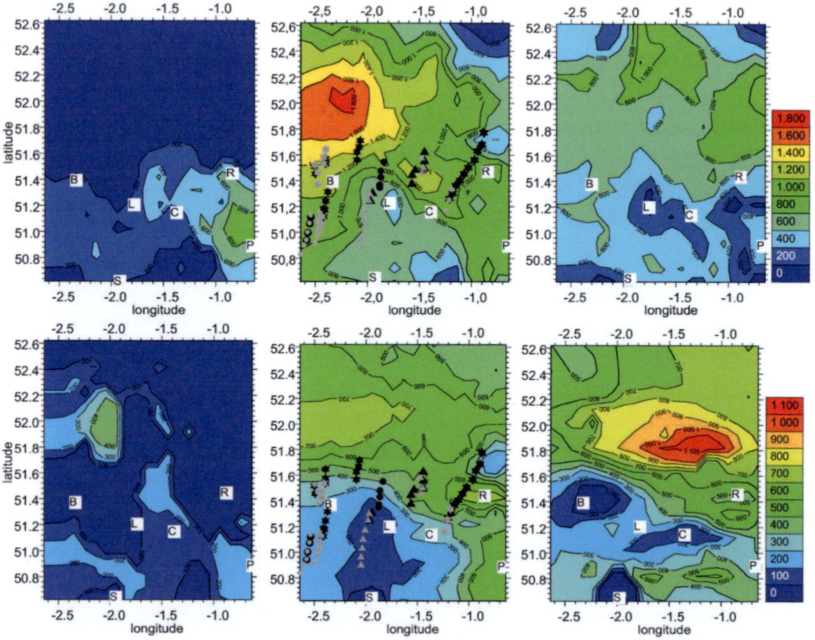

Figure 7.4: CAPE in J kg^{-1} at 1000 UTC (left), 1200 UTC (center) and 1400 UTC (right). Top) surface-based parcel method, and bottom) 50 hPa mixed-layer parcel method. The letters indicate the positions of the radiosondes. The cloud tracks are shown in grey and the radar tracks are plotted in black (referred to in the text). The cloud and radar tracks in the period from 1130 to 1145 UTC are shown as circles, at 1200 UTC as stars and at 1230 UTC as triangles.

layer at these sites is in agreement with the evolution near the surface except in the humidity distribution in the area of Bath (Tab.5.1). At noon, this region corresponds to an area of high moisture at the near-surface whereas the boundary layer is very dry relative to the distribution in the area. However, the behaviour of the humidity in the boundary layer corresponds with the IWV evolution, which shows a decrease in the areas where boundary layer drying is observed (Fig.6.7). The strongest temperature increase in the boundary layer is observed in the center of the CSIP domain, i.e. between Reading and Bath, while the strongest drying is detected at Larkhill and Chilbolton. The difference found between the surface observations, IWV (Fig.6.7) and boundary layer measurements at noon (Tab.5.1) in the area of Bath could explain the different patterns observed in the distribution of CAPE in this area when using the SB parcel or the 50 hPa ML parcel. The strong increase in CAPE from values of about 100 J kg^{-1} to 700 J kg^{-1} in the area of Reading between 1000 to 1200 UTC is remarkable. This significant increase in conditional instability responds to the strong increase in temperature and moisture in the PBL estimated for this station (Tab.5.1).

7.2.2 Distribution of CIN

Fig.7.5 shows that by 1000 UTC, the SB CIN field indicates low inhibition in the southern areas increasing towards the north. However, the CIN field based on 50 hPa ML parcels shows low inhibition in the west, about -10 J kg^{-1} in the area of Bath, and higher inhibition in the east, about -110 J kg^{-1}, in the area of Reading. By 1200 UTC, the CIN values based on 50 hPa ML data, are already higher than -15 J kg^{-1}. Noticeable is the strong reduction of inhibition at Reading from values of about -110 J kg^{-1} at 1000 UTC to values close to zero at 1200 UTC. In contrast to the 50 hPa ML parcel choice the SB parcel choice indicates some areas with higher inhibition in the northern and southern parts of the investigation area. Currently, maps from many organizations show contour CIN values at intervals of -50 J kg^{-1}. In agreement with our observations on this day, former publications suggested that smaller contour intervals are appropriate (Zieger et al., 1997). Due to the low CIN and high CAPE values in most parts of the CSIP domain at 1200 UTC, convection could be expected in many areas. However, convection is only initiated in a few places. Thus, the analysis suggests that other parameters besides CAPE and CIN are required when investigating initiation of deep convection. Using the 50 hPa ML parcel choice, the region around Bath shows low CAPE because of the low moisture content in the boundary layer. This could explain that the cells initiated in this area disappear shortly afterwards whereas the cell initiated between Chilbolton and Reading intensifies. Probably, in the area of Bath the initiation of convection occurs due to parcels lifted through buoyancy from the near-surface. For that reason, the SB parcel choice seems to be more representative of the conditions leading to initiation of convective cells while the use of the 50 hPa ML parcel better represents the likelihood of deep convection.

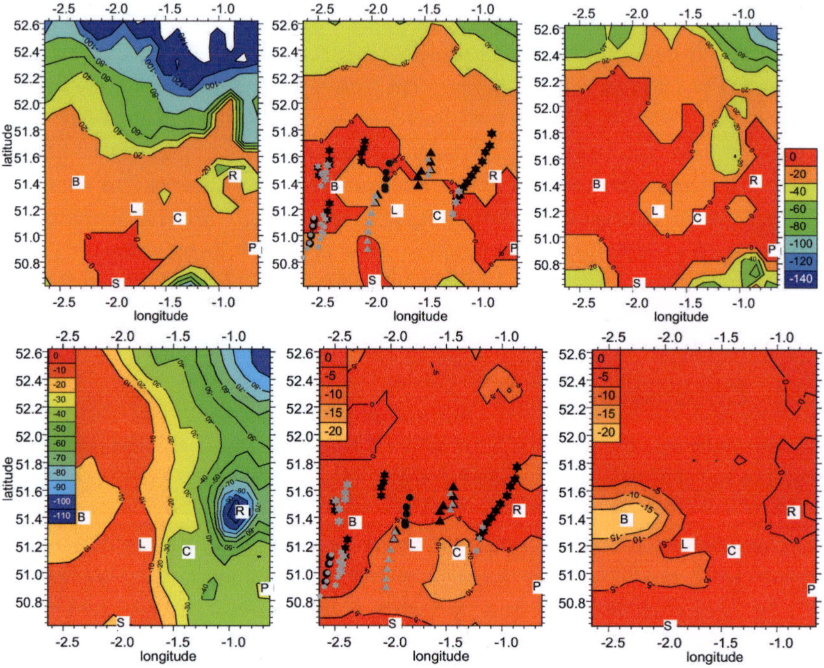

Figure 7.5: CIN in J kg^{-1} at 1000 UTC (left), 1200 UTC (center) and 1400 UTC (right). Top) surface-based parcel method, and bottom) 50 hPa mixed-layer parcel method. The letters indicate the positions of the radiosondes. The cloud tracks are shown in grey and the radar tracks are plotted in black (referred to in the text). The cloud and radar tracks in the period from 1130 to 1145 UTC are shown as circles, at 1200 UTC as stars and at 1230 UTC as triangles.

7.3 Analysis of additional parameters

As suggested in the last section, an analysis of additional convection-related indices, others than CAPE and CIN, is required when investigating initiation of convection. Especially, other parameters representing the inhibition of the lower atmosphere are needed because the low CIN in the area makes it difficult to observe the spatial differences. Representing the atmospheric stability, the LI and KO indices are analyzed. Representing the atmospheric inhibition, the CAP and lid effect indices are investigated. To ascertain the influence of wind shear, the BRN is investigated. Since no additional significant result is obtained from this analysis, the values obtained are not shown. Calculations are shown using the 50 hPa ML parcel choice. In addition, parameters representing the possible triggering in the area are presented.

7.3.1 Atmospheric stability

In general, the spatio-temporal evolution of LI (Fig.7.6a) follows similar patterns such as the ones observed on the 50 hPa ML CAPE fields (Fig.7.4). By 1000 UTC, CAPE is low in most parts of the investigation area. Only in the north-west, at about 52° N, maximum CAPE around 400 J kg^{-1} is observed. In this area, LI is about -1.5 °C. In agreement with the CAPE, LI shows low conditional instability in the areas of Larkhill, Chilbolton and Preston. Elsewhere the atmosphere is stable. The KO-index (Fig.7.6b) indicates moderate potential instability in the west. By 1200 UTC, the general increase in CAPE especially in the north and east of the investigation area is reflected in the LI fields. LI shows destabilization of the atmosphere in the whole area. Especially in the eastern and northern areas where maximum CAPE values prevail, high conditional instability, LI $\sim$ -2 to -4 °C, is observed. At this time, in the area of Reading and north of it, KO $\sim$ -4 K indicates that the atmosphere is potentially unstable. In the area of Reading from 1000 to 1200 UTC, CAPE increases from less than 100 J kg^{-1} to approximately 700 J kg^{-1}, LI also shows a strong destabilization in this period, LI $\sim$ +1 °C and KO $\sim$ + 1 K at 1000 UTC, to LI $\sim$ -5 °C, and KO $\sim$ -4.5 K at 1200 UTC. From 1200 to 1400 UTC, a significant increase in CAPE in a west-east oriented line at about 52° N is also reflected in the LI field. In this period, LI varies from -2.5 °C to -5 °C, indicating strong conditional instability in this area. The KO-index is about -1 to -2 K. In general, the three indices examined representing the atmospheric stability, CAPE, LI, and KO-index, show a similar spatio-temporal evolution.

7.3.2 Convective Inhibition: evolution of the stable layers

At 0800 UTC, all radiosondes show a strong lid between about 930 and 925 hPa ($\sim$ 500 m agl), capping a stably-stratified residual layer (not shown). By 1000 UTC, a well-mixed boundary layer which reached this capping inversion at all radiosonde

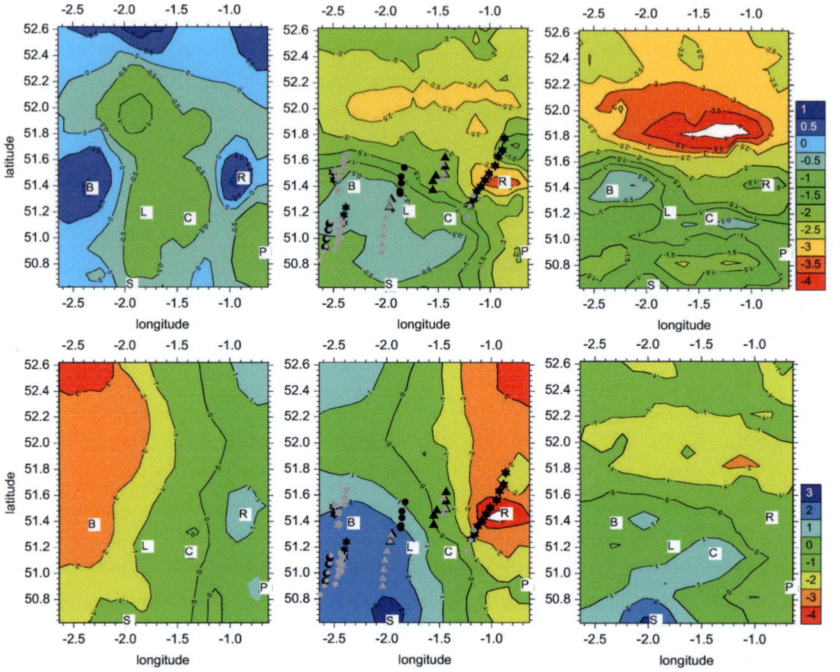

Figure 7.6: Spatio-temporal evolution of (top) LI in °C, and (bottom) KO-index in K, at 1000 UTC (left), 1200 UTC (center) and 1400 UTC (right). The letters indicate the positions of the radiosondes. The cloud tracks are shown in grey and the radar tracks are plotted in black (referred to in the text). The cloud and radar tracks in the period from 1130 to 1145 UTC are shown as circles, at 1200 UTC as stars and at 1230 UTC as triangles.

sites is observed. As an example, the temperature and dew point temperature from the radiosondes at Reading and Preston Farm at 1000 UTC and 1200 UTC are presented in Fig.7.7 to show the temporal evolution of lids at these sites. The soundings at 1000 UTC reveal the presence of multiple stable layers. They are labelled in Fig.7.7 as L1 to L3. Each of these layers has the potential for slowing the development of convection or even the formation of convective clouds, e.g. the strong low-level inversions (L1) centered at 930 hPa at Reading and at 925 hPa at Preston Farm. These low-level lids assist in the build-up of conditional instability by allowing warm, moist air to accumulate at low-levels, thereby increasing CAPE.

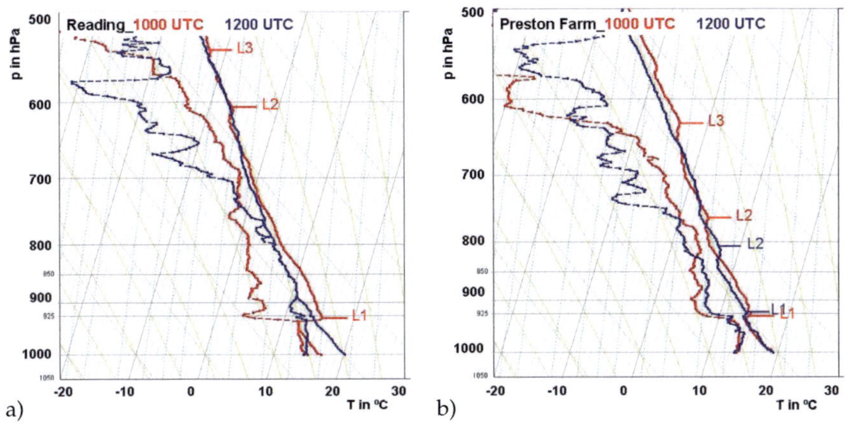

Figure 7.7: Skew T-log p diagram with temperature and dew-point temperature profiles at 1000 UTC (red) and 1200 UTC (blue) at a) Reading, and b) Preston Farm on 29 June 2005. L1, L2 and L3 indicate the positions of the lids referred to in the text.

Until 1200 UTC, surface heating leads to a strong warming in the boundary layer (Tab.5.1) contributing to the complete erosion of L1 at Reading. L2 and L3 do not exert any constraining effect anymore. In contrast to the evolution of the lids observed at Reading, at Preston Farm, less than 50 km to the south-east of Reading, L1 is still present since almost no change in the boundary layer temperature and moisture occurs at this station between 1000 UTC to 1200 UTC (Tab.5.1). The evolution of ML CAPE in the areas of Reading and Preston Farm shows that indeed the low-level lid, L1, at both stations contributes to the increase of CAPE in the boundary layer in the period from 1000 UTC to 1200 UTC, however, at 1200 UTC the erosion of L1 at Reading allows the release of this CAPE, whereas at Preston Farm the formation or development of convective clouds is still constrained (Fig.7.7).

The spatio-temporal evolution of the lids can be investigated using the CAP and lid effect indices (Fig.7.8a,b). The analysis of CAP is used to show the different spatial low-level inhibition conditions to determine the inhibition that a parcel should overcome to reach the LFC. The lid effect is used to analyze the distribution of the strongest stable layer between 850-500 hPa. Calculations are shown using the 50 hPa ML parcel choice. At 1000 UTC, in agreement with the distribution of CIN (Fig.7.5b), strong capping below the LFC is observed in the eastern areas, with CAP $\sim$ 3.5 °C at Reading and Preston Farm, whereas in the west CAP is about 1 °C. The spatial distribution of mid-tropospheric lids shows a similar pattern. However, the lid effect indicates that the strength of the strongest mid-level inversion is much higher at Preston Farm (lid effect $\sim$ 2 °C) than at Reading (lid effect $\sim$ 0.5 °C), as is also shown in Fig.7.7. At 1200 UTC, the low-level inversions observed at 1000 UTC have weakened in the whole area. The reduction of the inversion strength has decreased at the different stations and is highly related with the corresponding boundary layer warming (Tab.5.1). This was especially noticeable in the area of Reading where the CAP $\sim$ 3.5 °C is reduced to CAP $\sim$ 0 to 0.25 °C. In the south, the rest of the stations show CAP $\sim$ 0.5 to 1 °C, whereas in the north CAP is in the order of 0 to 0.5 °C. In particular, at Preston Farm the strong low-level inversion observed at 1000 UTC is still present at 1200 UTC (Fig.7.7) because of the little warming observed at this site (Tab.5.1). At this time, mid-tropospheric lids no longer exert a constraining effect in the area of Reading and in the northern areas; however, at Preston Farm, Swanage and Chilbolton a mid-tropospheric inversion is still present (Fig.7.8). At 1400 UTC in the area of Bath a low-level inversion, CAP $\sim$ 1 °C, and a mid-tropospheric inversion, lid effect $\sim$ 1 °C, exists. At Preston Farm, Swanage and Chilbolton weak low-level and mid-tropospheric inversions are present at this time.

From the distribution of CAP and lid effect and the observed development of convection one may conclude that at 1200 UTC the initiation of convection takes place in the transition zones between strong and weak capping conditions. In particular, the cell that breaks out close to Chilbolton intensifies when advected towards the north into an area where a strong inversion in the morning is present, which leads to a significant increase of temperature and moisture in the boundary layer, eventually removing the low-level inversion. Moreover, the line of thunderstorms observed at 1500 UTC organizes in an area where no low-level or mid-tropospheric inversions exist after 1200 UTC, i.e. the formation and development of convective clouds is not constrained in any way.

7.3.3 Triggering

By 1000 UTC, in the northern areas Tc-Tv $>$ 3 °C. Therefore, Tc in these areas is far from being reached (Fig.7.9). However, in the CSIP domain the difference is much smaller. Tc is reached in the area of Bath and Swanage at this time. At 1200 UTC, the area where the Tc-Tv$\leq$0 condition is fulfilled is larger, but remains close to the area of Bath and the area of Reading. By 1400 UTC, the area where Tc is reached is even larger, but still there are areas where Tc-Tv $>$ 0. This means that in the northern areas where high instability (Fig.7.4) and low inhibition (Fig.7.5) is observed, an

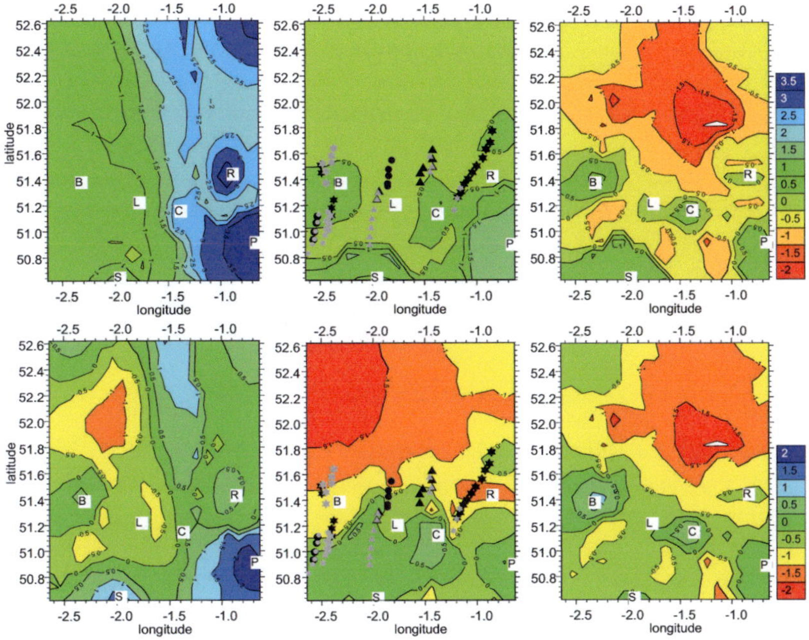

Figure 7.8: Spatio-temporal evolution of top) CAP in °C, and bottom) lid effect in °C, at 1000 UTC (left), 1200 UTC (center) and 1400 UTC (right). The letters indicate the positions of the radiosondes. The cloud tracks are shown in grey and the radar tracks are plotted in black (referred to in the text). The cloud and radar tracks in the period from 1130 to 1145 UTC are shown as circles, at 1200 UTC as stars and at 1230 UTC as triangles.

additional triggering mechanism is needed for the initiation of convection. At 1000 UTC, near-surface horizontal convergence is observed in several areas (Fig.7.9). In particular, stronger convergence in the CSIP domain, around -0.15 x 10^{-3} s^{-1}, took place in the area between Larkhill and Chilbolton both at 1000 and 1200 UTC. By 1200 UTC, strong convergence is observed also along a west-east oriented line at about 52° N. At this time, below 52° N, south and south-westerly winds are observed whereas above this latitude mainly a westerly flow occurs. The confluence of both wind flows at about 52° N results in the convergence line observed. By 1400 UTC, the convergence in this line has increased to about -0.2 x 10^{-3} s^{-1}. The calculation of the boundary layer horizontal wind convergence is in this case limited to the area where radiosondes are available. Therefore, the investigation area is restricted from 50.62° N to 51.42° N and from 2.34° W to 0.74° W. The wind field used to estimate the PBL horizontal divergence is based only on radiosonde data. By 1000 UTC, the strongest horizontal convergence of about -0.04 x 10^{-3} s^{-1} is observed in the south of Reading. By 1200 UTC, an area of horizontal convergence, approximately -0.06 x 10^{-3} s^{-1} is observed between Reading and Chilbolton. By 1400 UTC, the strongest convergence, $\sim$ -0.02 x 10^{-3} s^{-1}, is in the area of Bath. Divergence fields from the PBL and near-surface horizontal wind generally agree, although in the higher resolution divergence field the convergence lines appear narrower.

7.4 Deep convection likelihood

For predicting the likelihood of deep convection, commonly sounding-derived parameters, from real or forecast soundings, are used and plotted in horizontal charts. None of the convection-related indices investigated in the last section can be used to predict the initiation of deep convection on its own. For example, not all convergence lines initiate storms, even when they collide in apparently conditionally unstable environments (Stensrud and Maddox, 1988), and very low CIN does not always indicate that there is a high likelihood of deep convection. In atmospheric situations with very low inhibition, this may lead to the continuous release of latent and sensible heat as small clouds, preventing sufficient energy build-up to drive strong thunderstorms. Therefore, the convection-related parameters investigated must be evaluated jointly to ascertain the likelihood of deep convection (Weckwerth, 2000; Graziano and Carlson, 1987).

The overall aim in this section is to find an optimal combination of convection-related indices to describe convection initiation. The convection-related indices investigated in the last section are combined and jointly evaluated to ascertain the deep convection likelihood for the CSIP IOP5. The working hypothesis is that areas where deep convection occurs must possess conditional instability, indicated by high CAPE and low relative LI, low atmospheric inhibition, indicated by low relative CIN and CAP below 2 °C and absence of mid-troposopheric lids, i.e the lid effect index must be close to zero. Moreover, a triggering mechanism must exist.

7.4.1 The need of jointly evaluating the convection-related indices

In Fig.7.10, different combinations of the convection-related indices used in the last section are presented. The KO-index has been excluded in this analysis because

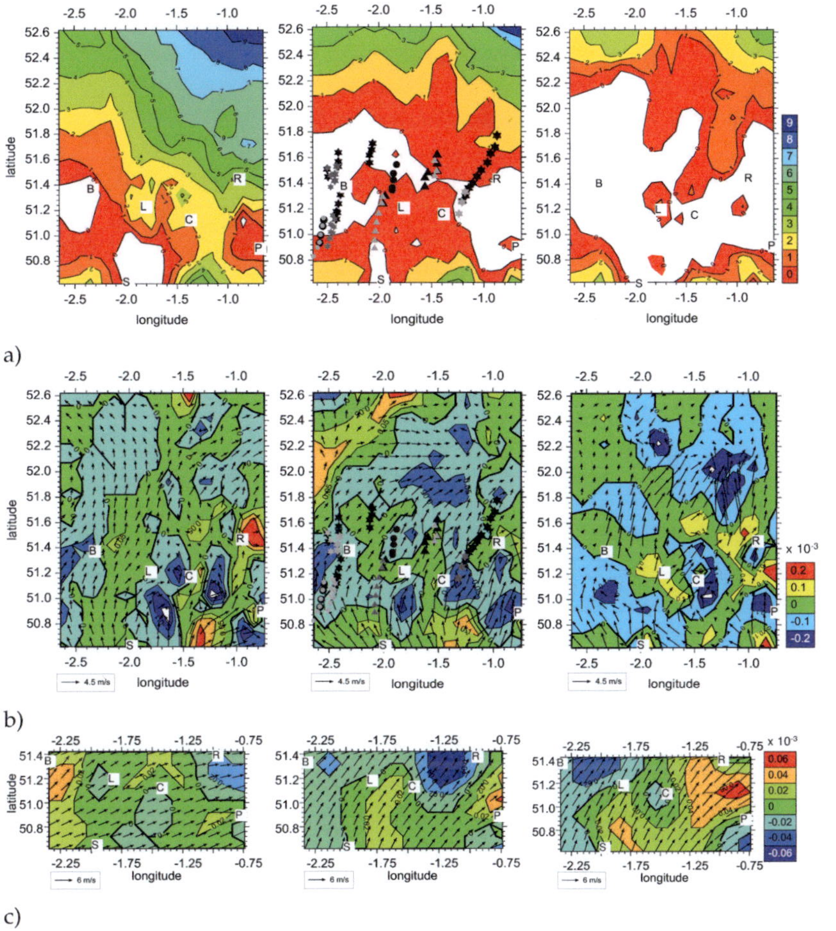

Figure 7.9: Spatio-temporal evolution of a) Tc-Tv in °C, b) the near-surface horizontal divergence field, and c) PBL horizontal divergence field x 10^{-3} s^{-1} (colour code) and interpolated wind field in m s^{-1} (wind vectors), at 1000 UTC (left), 1200 UTC (center) and 1400 UTC (right). The letters indicate the positions of the radiosondes. The areas in white in the Tc-Tv fields indicate Tc-Tv < 0. The cloud tracks are shown in grey and the radar tracks are plotted in black (referred to in the text). The cloud and radar tracks in the period from 1130 to 1145 UTC are shown as circles, at 1200 UTC as stars and at 1230 UTC as triangles.

synoptic scale forcing was very weak in the case of the day discussed. From now on, this index will be used only on the days when large-scale forcing plays any role in the initiation of deep convection. To illustrate the discussion, the calculated indices for three different radiosondes in the area are highlighted.

In the first case, a radiosonde close (to the west) of the Reading radiosonde station is used and it is indicated by a full green circle. High conditional instability is observed, CAPE $\sim$ 700 J kg^{-1} and LI $\sim$ -2.5 °C. Also, there is very low inhibition, indicated by CIN $\sim$ -1 J kg^{-1} and CAP $\sim$ 0.2 °C. Moreover, buoyancy (B) and lid effect are within the thresholds where deep convection could be expected (Graziano and Carlson, 1987), B $<$ 1 °C and lid effect $<$ 2 °C, since in this case, B $\sim$ -0.5 °C and lid effect $\sim$ -1.5 °C. However, the fourth diagram indicates that the convective triggering temperature is not reached, Tc-Tv $>$ 0, therefore, parcels are not triggered by near-surface buoyancy. Moreover, no convergence is observed in this area. Although the observed stability and inhibition are favourable for high deep convection likelihood, the absence of a triggering mechanism prevents initiation.

A second case, a radiosonde close to Bath radiosonde station is chosen and it is indicated by a full blue circle. It is selected because Tc is reached at the near surface, but deep convection is not observed. In this case, one may observe that CAPE is very low, CAPE $\sim$ 200 J kg^{-1}, and LI is close to zero. However, CIN is low, CIN $\sim$ -5 J kg^{-1} and CAP $\sim$ 0.6 °C, but higher than in the former case. B is slightly below 1 °C and the lid effect $\sim$ 0.5 °C. Therefore, despite the existence of a triggering mechanism, deep convection likelihood is reduced since the CAPE is very low.

The third case, a radiosonde to the north-east of the Chilbolton radiosonde station is used and it is indicated by a full black circle. This is an area where initiation of convection was observed. In this case, moderate CAPE $\sim$ 400-500 J kg^{-1} and LI $\sim$ -2 °C are observed. CIN $\sim$ -4 J kg^{-1} and CAP $\sim$ 0.5 °C indicate low inhibition in the area. Furthermore, B $\sim$ -0.5 °C and lid effect $\sim$ -0.5 °C indicate that cells can reach high levels without any constraining effect. In addition, both Tc-Tv $<$ 0 and the near-surface convergence indicate that parcels from the near surface can freely rise and additional lift by wind convergence will help these parcels to reach their LFC and form deep convective clouds. In this case, all necessary conditions for the initiation of deep convection are fulfilled, and deep convection would be expected, in agreement with the reality.

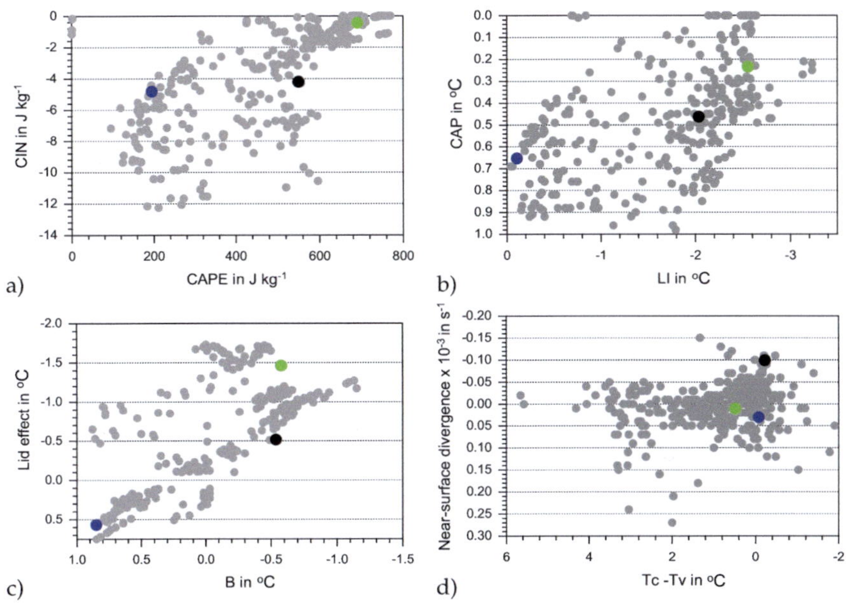

Figure 7.10: 2-D plots of convection-related indices at 1200 UTC on 29 June 2005. a) CAPE vs CIN, b) CAP vs LI, c) lid effect vs B, and d) near-surface divergence vs Tc-Tv. The grey points indicate the results obtained for each grid point in the spatial distribution fields of convection-related indices. The blue, green and black points indicate different atmospheric situations. A detailed description of the latter is found in the text.

7.4.2 Deep convection likelihood based on the combination of gridded convection-related indices

Up to now different convection indices were analyzed to quantify the required conditions for the occurrence of deep convection. In this section, the information given by the different convection indices is jointly evaluated in combination with the parameters that represent the possible triggering in the area, to ascertain whether such combination and the resolution obtained by the combination of radiosonde and GPS measurements are enough to improve the representation of the location of the initiation of deep convection.

The following convection-related indices are discussed: a) CAPE and CIN, b) LSI, c) the Tc-Tv difference and d) the near-surface divergence field on 29 June at 1200

UTC (Fig.7.11). CAPE, CIN and the LSI are computed using the 50 hPa ML parcel whereas Tc-Tv is calculated using the near-surface data. Until now, two indices have been used to describe low-level and mid-tropospheric inversions, CAP and lid effect, respectively. When dealing with the initiation of convective cells it is useful to investigate separately these two parameters. A low-level inversion, below the LFC, suppresses convection whereas a mid-tropospheric lid only constrains its evolution in the atmosphere. In this section, however, the lid effect, term B in the LSI index (Eq. 2.31), is computed between 950 and 500 hPa to include the inhibitor effect of the low-level inversions, since we are not only interested in the initiation but also in the intensification of deep convection. That means, neither strong lids should exist below the LFC nor strong mid-tropospheric inversions should be present.

From all the initiated cells observed in the period from 1130 to 1230 UTC, the one that initiates at 1200 UTC in the south-east between Chilbolton and Reading has the longest life and produces the strongest showers (Fig.7.2). All cells initiate in areas of low to moderate CAPE, especially the ones in the area of Bath where CAPE was only 200 J kg^{-1} (Fig.7.11a). North-east of Chilbolton CAPE is about 400 J kg^{-1} increasing towards Reading where CAPE is approximately 700 J kg^{-1}. This means that once convection is initiated, the amount of energy to be released is much higher in that cell than in the others, leading to stronger showers. The increase in CAPE and LSI (Fig.7.11b) in the surrounding of Reading is intimately related with the evolution of temperature and moisture in the PBL. While in the area of Reading, strong warming and moistening is observed in the PBL between 1000 to 1200 UTC, in the area of Bath and Larkhill little warming and drying take place (Tab.5.1). Probably, the high surface temperature and moisture in the area of Bath (Fig.6.3) results in Tc-Tv ≤ 0 so that thermals could be lifted by surface-triggered buoyancy in this area. But, the progressing mixing of the parcels with the dry and cold boundary layer air (Mahrt, 1976) observed in this area (Tab.5.1) results in a loss of buoyancy which, in addition to the low CAPE, leads to a rapid decay.

Initiation of convection seems to be highly related with parcels being lifted from the surface due to buoyancy effects since most of the convective cells initiate in areas where Tc is reached (Fig.7.11c). CIN is very low ($>$ -15 J kg^{-1}) in the whole area so that only a small trigger mechanism is necessary for the parcels to reach their LFC. For example, the convergence of -0.15 x 10^{-3} s^{-1} observed between Chilbolton and Reading where the cell breaks out (Fig.7.11d), could have led to stronger updrafts which overcame the low CIN of -10 J kg^{-1}. An additional area with relatively high

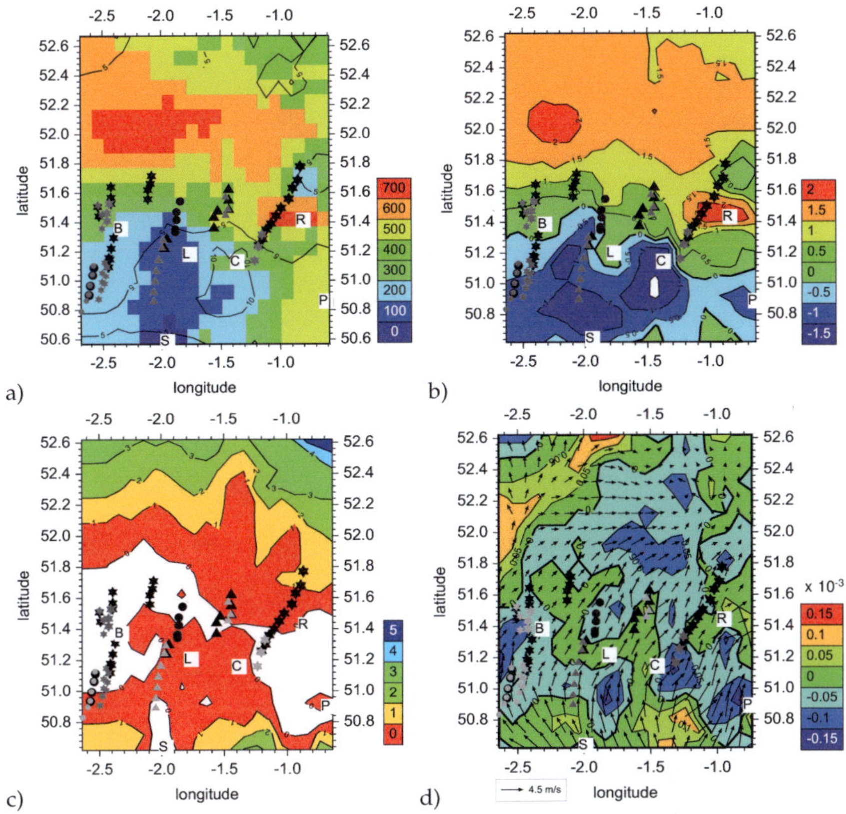

Figure 7.11: Spatial distribution of the convection-related indices at 1200 UTC on 29 June 2005. a) CAPE (colour scale) and CIN (isolines) in J kg^{-1}, b) LSI in °, c) Tc-Tv in °, and d) near-surface divergence field x 10^{-3} s^{-1} (colour scale) superimposed are the horizontal wind vectors in m s^{-1}. The letters indicate the positions of the radiosondes. The cloud tracks are shown in grey and the radar tracks are plotted in black (referred to in the text). The cloud and radar tracks in the period from 1130 to 1145 UTC are shown as circles, at 1200 UTC as stars and at 1230 UTC as triangles. The areas in white in the Tc-Tv field indicate Tc-Tv < 0.

CAPE (600 to 700 J kg^{-1}), high LSI (about 1.5° C), convergence (-0.1 to -0.15 x 10^{-3} s^{-1}) and almost no CIN (about -5 J kg^{-1}) is observed along an east-west oriented line at about 52° N (Fig.7.11a, b, d). Until 1400 UTC, CAPE in this region increases to about 1100 J kg^{-1} (Fig.7.4) and the convergence to approximately -0.3 x 10^{-3} s^{-1}

(Fig.7.9). Although no initiation is observed at 1200 UTC, this region corresponds to the location of the well-organized line of thunderstorms at 1500 UTC accompanied by heavy precipitation (Fig.7.3). Thus, the convection-related indices correctly indicate that the likelihood of deep convection is high in the area north of Reading.

From this investigation it could be concluded that in this event storm initiation depend more on the near-surface temperature, boundary-layer convergence and low CIN than on high CAPE values. Location and timing of the initiation of convection are critically influenced by the structure of the humidity field in the planetary boundary-layer. Additionally, the initiation is better represented by indices calculated with a parcel lifted from the near-surface whereas the areas where deep convection intensifies is better detected by indices calculated with parcels characterized by mixed-layer conditions. Once triggered, the presence of moderate or high CAPE and a high amount of mid-troposphere moisture helps initiated convective cells to grow into deep convection. In the next chapter, different IOPs are investigated to ascertain the usefulness of this combination of convection-related indices and the application of the adjustment method to predict deep convection likelihood in different atmospheric situations.

Chapter 8

Case Studies

In this chapter, different IOPs from the CSIP campaign; IOP2, IOP8, and IOP16 and one from the COPS campaign, IOP8b, are investigated. In all cases, initiation of convection is due to boundary layer processes. Within the boundary layer a wide range of processes may be responsible for the initiation of convection in the area. Each IOP is characterized by a distinctive feature that controls the convection initiation and evolution. Due to the different processes dominating in each case, the highlights are different. Moreover, a case from the COPS campaign, complex terrain, is analyzed and compared with a similar case from the CSIP campaign, flat terrain, to find the similarities and differences observed in the processes that lead to the initiation of isolated storms in both types of topography. The main goal in this chapter is to demonstrate that an increase in resolution at the near-surface and in the boundary layer helps explaining the processes leading to the initiation

	June 19, 2005 CSIP IOP2	July 13,2005 CSIP IOP8	August 18, 2005 CSIP IOP16	July 15, 2007 COPS IOP8b
Primary initiation	NO	YES	YES	YES
Intense showers	NO	YES	YES	YES
Hill effects	NO	YES	YES	YES
Coastal effects	NO	YES	NO	NO
Convergence lines	YES	YES	YES	YES
Existence of lids	YES	YES	YES	YES
Multiple lids	NO	YES	NO	NO
Main lid height varies in space	NO	NO	YES	NO

Table 8.1: Summary of the observations in the investigation domain for the IOPs analyzed in this chapter.

of deep convection and the combination of convection-related indices proposed in the former chapter works properly under different atmospheric situations in order to explain initiation of convection forced by boundary layer processes. In Tab.8.1, a summary of the main observations in the investigation domain on the IOPs investigated is presented.

8.1 CSIP IOP2: Suppressed deep convection

In the IOP2, 19 June 2005, deep convection and heavy showers are observed in the west of UK. In the CSIP domain, thermals develop in the boundary layer in the morning, particularly to the west of Chilbolton and eventually a few cumulus showed up, but they did not develop further, therefore, this event is named "suppressed deep convection". The analysis of this day is devoted to investigating why despite the high instability observed in the CSIP area no deep convection occurs.

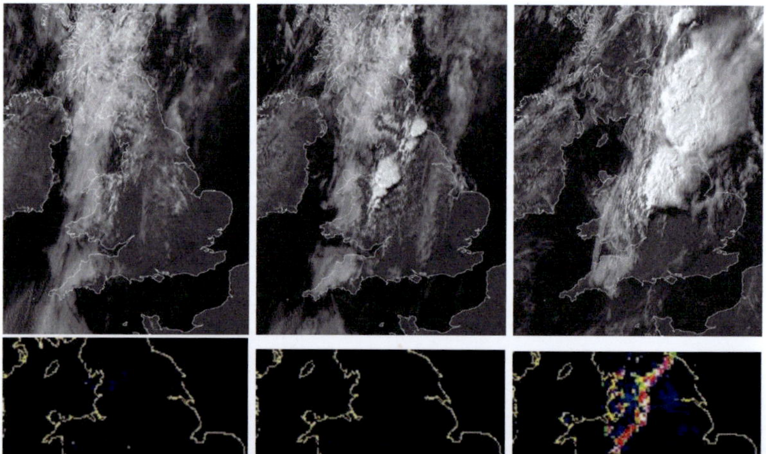

Figure 8.1: MSG visible imagery (top) and rain radar (bottom) at 1000 UTC (left), 1200 UTC (center), and 1600 UTC (right) on 19 June 2005.

8.1.1 Synoptic situation

The analysis of the mean-sea-level pressure shows a cold front that remained slow-moving to the west of the CSIP area (Fig.8.2). By this time, large-scale forcing with lifting of 0 to -10 hPa h^{-1} is observed along a SW-NE oriented band in the west of the UK. In the same area, large potential instability, KO $\sim$ -12 K, is present. In the CSIP area, negative vorticity advection in upper levels and weak warm air advection in the layer from 1000 to 500 hPa result in subsidence. The Global Forecast System (GFS) model 10m wind (not shown) shows south-south-westerly winds covering most of the UK area, except in the west where westerly winds are observed in association with the cold front moving from the west. The confluence of these winds is related to a convergence zone (frontal convergence) in a north-south oriented line which may have been the reason for the initiation of convection in this area. At 500 hPa, a south-westerly flow is observed in the whole area.

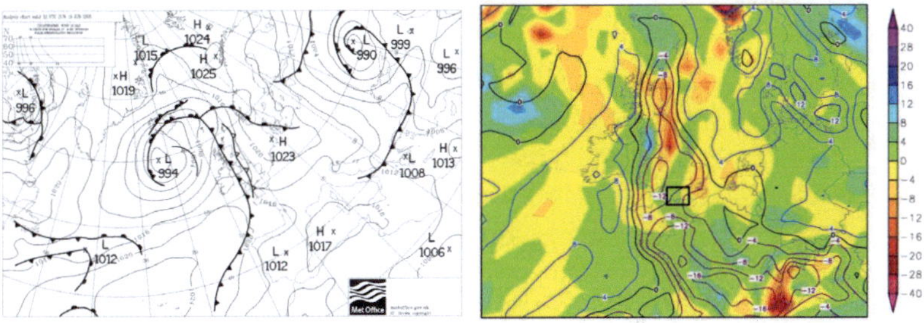

Figure 8.2: Met Office analysis of sea level pressure (left) and GFS analysis of omega in hPa h^{-1}(colour scale) at 500 hPa and KO-index in K (isolines) (right), at 1200 UTC on 19 June 2005.

8.1.2 GPS observations of the atmospheric water vapour

Ahead of the cold front an area of high moisture is observed extending from south to north (Fig.8.3). The moisture gradient from east to west was already present at 0500 UTC. The south-westerly flow slowly advects the moist air mass towards the north-east. A moisture gradient of about 7 to 10 kg m^{-2} between the CSIP and the north-western area is present through the whole day (Fig.8.3). In the south-eastern region, dry air is advected from the south which makes the gradient from south-east to north-west even larger through the day. In particular, in the CSIP area (Fig.8.4), an east to west IWV gradient is found in agreement with the general pattern observed in Fig.8.3. By 1000 UTC, IWV values are low in the eastern region,

$\sim$ 30.5 kg m^{-2}, and high in the north-west, $\sim$ 32.5-33.5 kg m^{-2}. From 1000 to 1400 UTC, IWV is reduced in the whole area, especially in the eastern region it drops to about 27.5 kg m^{-2} at 1400 UTC. The IWV values observed in this day in the whole CSIP domain are comparable to or in some areas even higher than the observations on the IOP5, 29 June 2005, in areas where deep convection took place (Fig.6.7). On the IOP5, in the areas where high IWV was present also a wetter boundary layer with respect to the rest of the stations was observed. In order to investigate whether this pattern is reproduced on the IOP2, the spatio-temporal evolution of temperature and moisture at the near-surface and in the PBL is analyzed in the next section.

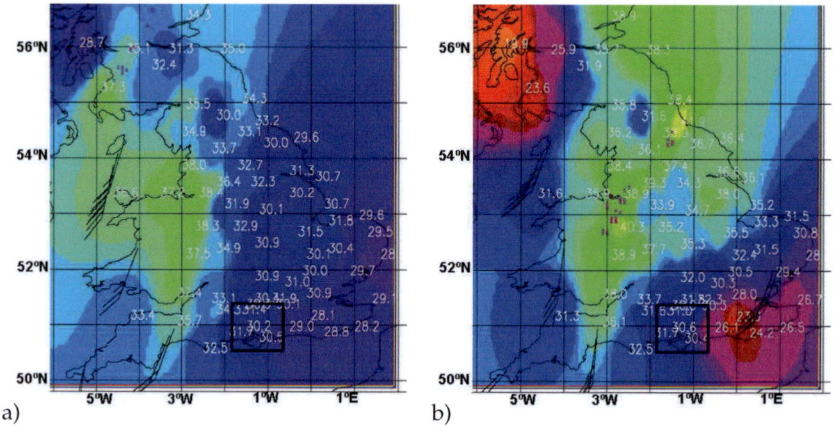

Figure 8.3: GPS-derived IWV in kg m^{-2} and 2 km winds at a) 0800 UTC, and b) 1400 UTC on 19 June 2005. The black box indicates the approximate location of the CSIP domain. A more detailed representation of the IWV-distribution in the CSIP domain is given in Fig.8.4.

8.1.3 Near-surface and boundary-layer distribution of temperature, moisture and wind field

For the analysis of the near-surface distribution of temperature, moisture and wind, AWSs and SYNOP measurements are used. The area of investigation extends from 50.62° N to 52.92° N and 0.74° W to 3.54° W. The CSIP domain of investigation is marked by a rectangle (Fig.8.5).

For the whole period investigated, from 1000 to 1400 UTC, an area of high temperature, about 17 °C to 20 °C, extends from south to north in the western region, at about 2.5° W. Temperatures in the east, especially in the CSIP domain (within the rectangle), are lower, about 15 °C at 1000 UTC increasing by about 2 °C from 1000 to 1400 UTC. The moisture distribution shows a different pattern. For the same time period, maximum mixing ratio is observed in the east, about 8 to 10 g kg^{-1}, and minimum in the west, from 6 to 8 g kg^{-1}. At 2 m level, a north-westerly flow is present in most of the area. In the CSIP area, south-south-westerly winds can also be seen at 1200 UTC. The north-westerly winds advect warm and dry air into the CSIP area through the day. In this case, the near-surface spatial distribution of moisture does not agree with the IWV pattern observed in Figs.8.3, 8.4. Whereas in general the IWV spatial distribution shows higher water vapour content in the west and lower in the east, the near-surface observations show wetter near-surface air in the east and drier air in the west.

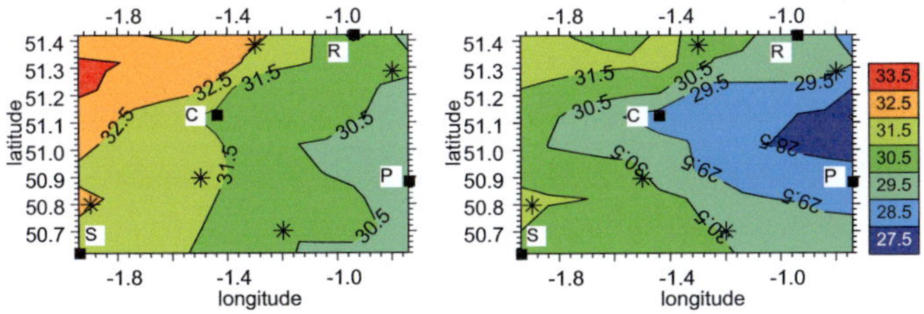

Figure 8.4: Interpolated fields of GPS-derived IWV in kg m^{-2} and IWV calculated from radiosondes in the area of investigation at 1000 and 1400 UTC on 19 June 2005. For the interpolation, GPS measurements outside of the CSIP domain are used. GPS stations are indicated by black stars, and radiosonde stations are indicated by black squares.

Radiosondes at Reading, Preston Farm, Chilbolton and Swanage are available on this day. The locations of the radiosonde stations in the CSIP domain are indicated in Fig.8.4. As an example, the evolution from 1000 to 1400 UTC at Chilbolton and Reading is shown in Fig.8.6. Little warming in the boundary layer and a very dry lower troposphere with shallow inversions at about 850 hPa and 600 hPa are observed. The boundary layer water vapour content remains constant in the period from 1000 to 1400 UTC; e.g for Chilbolton, the 50 hPa mean mixing ratio is about 11.1 g kg^{-1} at 1000 UTC, 10.9 g kg^{-1} at 1200 UTC, and 11.0 g kg^{-1} at 1400 UTC. At Reading, the temperature evolution is very similar to the one observed in Chilbolton. Boundary layer moisture content is approximately 11.3 g kg^{-1} at 1000 UTC, 10.5 g kg^{-1} at 1200 UTC and 10.0 g kg^{-1} at 1400 UTC. Thus, at 1400 UTC, the boundary layer at Reading is about 1.0 g kg^{-1} drier than at Chilbolton. Watnall

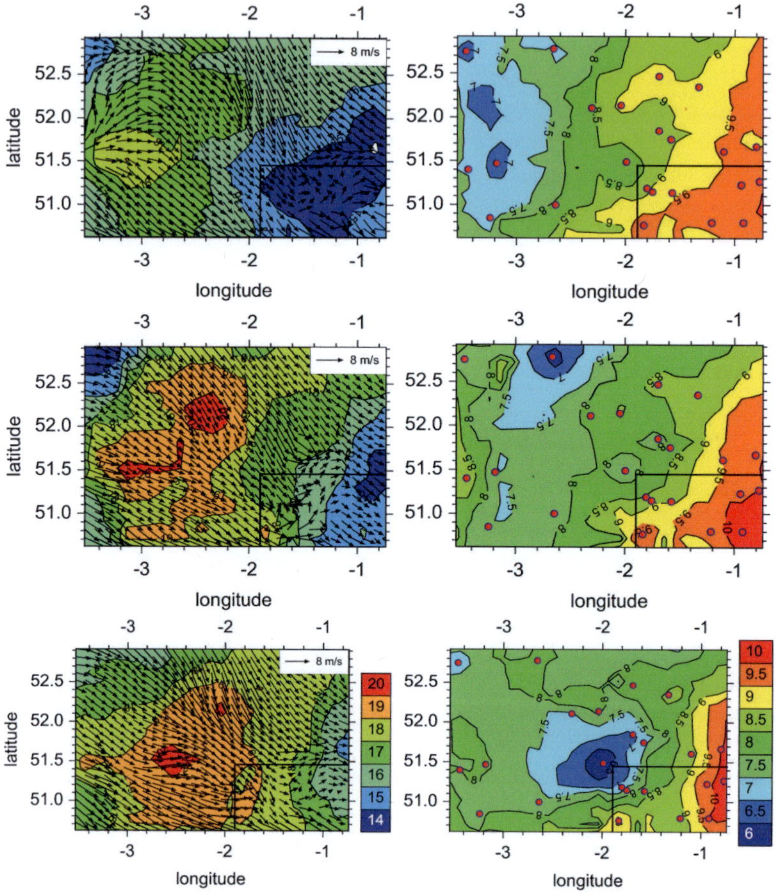

Figure 8.5: Spatio-temporal evolution of the near-surface temperature in °C (left), and mixing ratio in g kg^{-1} (right), at 1000 (top), 1200 (center) and 1400 UTC (bottom) on 19 June 2005. The CSIP investigation area is marked by a box. The red dots indicate the positions of the AWSs and the SYNOP stations. 2 m wind vectors are superimposed.

is an additional radiosonde station where a profile is available only at 1115 UTC during the day. It is located at about 53° N and 1.25° W. It shows the conditions in the north where high IWV values are observed. A comparison between the radiosondes at Reading and Chilbolton at 1200 UTC and the radiosonde at Watnall at 1115 UTC is presented in Fig.8.7. Little difference is found between the lower boundary layer temperature in the three radiosondes, however, at Watnall the troposphere is much wetter than those at Reading or Chilbolton, especially below 900 hPa and in a shallow layer between ~ 650 hPa and 550 hPa. The lower PBL distribution of moisture agrees reasonably well with the general pattern observed from the GPS IWV measurements of a wetter atmosphere on the western areas and a drier atmosphere in the south.

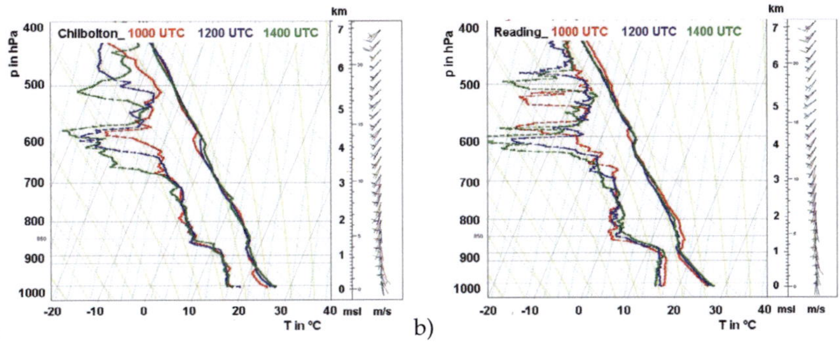

a) b)

Figure 8.6: Temporal evolution of radiosonde profiles at a) Chilbolton, and b) Reading, at 1000 (red), 1200 (blue), and 1400 UTC (green) on 19 June 2005. Wind bars in m s^{-1} (half-barb=2.5 m s^{-1}; whole-barb=5 m s^{-1}).

8.1.4 Convective indices

The difference in boundary layer water vapour content between the different radiosondes (Fig.8.7) largely influences the convection indices. Convection indices calculated at Watnall, Chilbolton and Reading applying the ML 50 hPa parcel method are summarized in Tab.8.2. The wetter PBL at Watnall results in higher atmospheric instability, as indicated by CAPE and LI, and lower inhibition, as indicated by CIN and CAP.

If at Reading we use the same mean boundary layer 50 hPa moisture measured at Watnall, which results in an increase of about 1 g kg^{-1}, not modifying the conditions above, the resulting convection indices are as follows: CAPE is about 1186 J kg^{-1}, LI ~ -4.6 °C, CIN ~ -14 J kg^{-1} and CAP ~ 3.0 °C, which implies a strong increase in instability and a reduction in inhibition. Making use of Eq.6.11, and using the radiosonde at Watnall as reference, one may estimate the boundary layer moisture content in the area where deep convection breaks out. Applying the relation between the boundary layer mixing ratio and the column-integrated water vapour at Watnall, $\sim$ 12.39 g kg^{-1} and 35.0 kg m^{-2}, respectively, and the IWV in the area where deep convection occurs, about 39.5 kg m^{-2}, we obtain a boundary layer water vapour content in the order of 14.0 g kg^{-1} for the area where deep convection takes place. If we use this representative boundary layer moisture in the PBL at Watnall without modifing the rest of the profile, the convection indices calculated are as follows: CAPE is about 2056 J kg^{-1}, LI $\sim$ -6.5 °C, CIN $\sim$ 0 kg^{-1} and CAP $\sim$ 0 °C. Thus, the conditions for the occurrence of deep moist convection would be ideal in the presence of the frontal convergence zone acting as a triggering mechanism. This is a clear example of how a difference of approximately 1 g kg^{-1} in the lower PBL results in more favourable conditions for the occurrence of deep convection. This result points out the importance of having an accurate representation of PBL water vapour to predict deep convection likelihood.

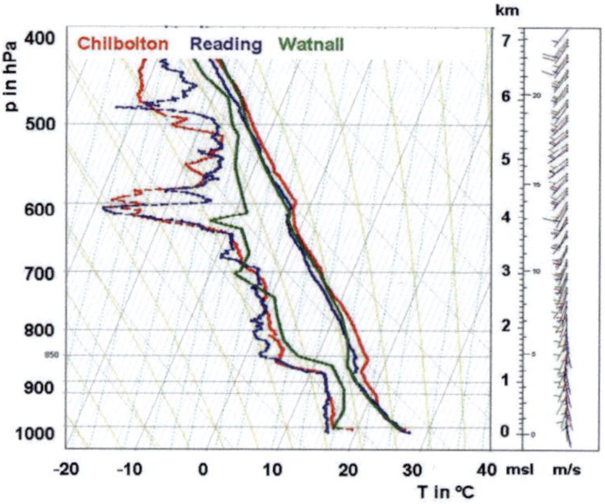

Figure 8.7: Comparison between radiosonde profiles at Watnall (green) at 1115 UTC, Reading (blue) at 1200 UTC, and Chilbolton (red) at 1200 UTC on 19 June 2005. Wind barbs in m s^{-1} are shown on the right of the diagram (half-barb=2.5 m s^{-1}; whole-barb=5 m s^{-1}).

Radiosonde station	time (UTC)	CAPE in J kg^{-1}	LI in °C	CIN in J kg^{-1}	CAP in °C
Watnall	1115	1150	-4.4	-13	2.0
Chilbolton	1200	640	-2.8	-140	5.2
Reading	1200	620	-3.0	-54	3.9

Table 8.2: Convection indices at Watnall, Chilbolton, and Reading applying the ML 50 hPa parcel choice.

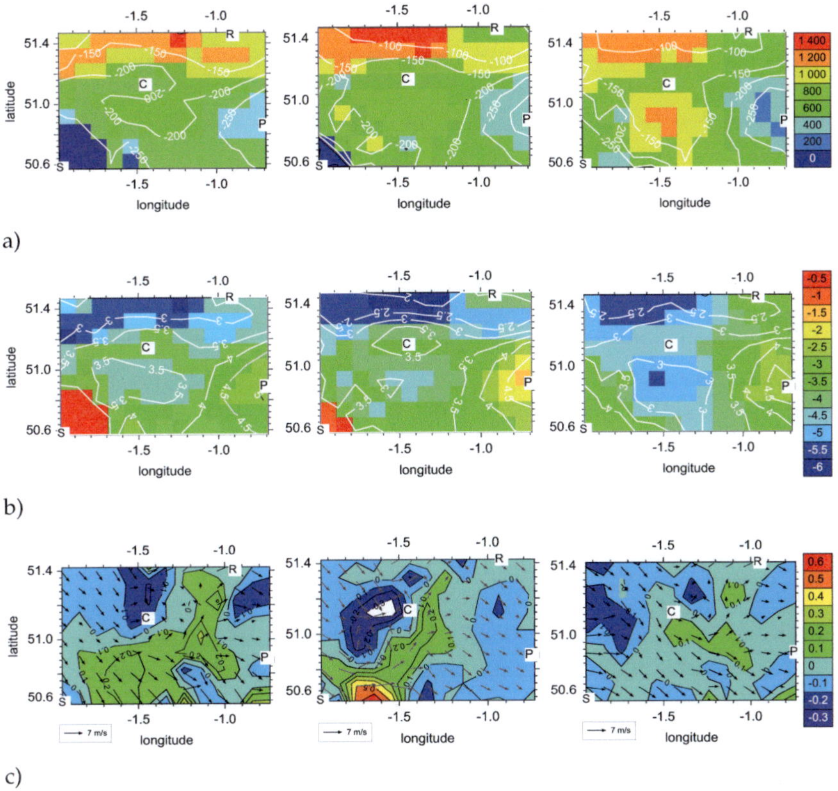

Figure 8.8: Combination of convection-related indices on 19 June 2005, at 1000 UTC (left), 1200 UTC (center) and 1400 UTC (right). a) CAPE in J kg^{-1} (colour code) and CIN in J kg^{-1} (isolines), b) LI in °C (colour code) and CAP in °C (isolines), and c) horizontal near-surface wind divergence x 10^{-3} s^{-1} in (colour code), and wind field in m s^{-1} (wind vectors). Convection indices are calculated using the 50 hPa mixed-layer parcel method.

Applying the adjustment method, the GPS IWV measurements are used to increase the resolution of the radiosonde-interpolated water vapour field. From the higher-resolution field, the spatial distribution of convection indices is calculated in the area of investigation from 1000 to 1400 UTC (Fig.8.8). Near-surface divergence fields are calculated from the high resolution AWSs network. As a first approximation, the high CAPE $\geq$ 1000 J kg^{-1}, and high negative LI < -4.0 °C, indicating large conditional instability in the northern areas during the day, could indicate a possible area for deep convection. However, the atmospheric inhibition in the area is very high. Although in the northern areas CIN and CAP show the lower absolute values, between -50 to -150 J kg^{-1} and between 2.5 to 3 °C, respectively, these are still high. The lower-level PBL-capping inversion at about 850 hPa (Fig.8.6) is not eroded during the whole time period as indicated by the spatial distribution of CAP in the area. The CIN values in the area (Fig.8.8a) normally indicate that the LFC can only be reached when triggering is present, because the equivalent vertical updraughts to overcome these values of CIN are about w $\geq(2CIN)^{0.5}$ = 10-32 m s^{-1}. Therefore, strong convergence would be necessary for parcels to break through the inhibition layer. Although strong convergence - relative to the observations on other days - is observed in some areas, -0.3 x 10^{-3} s^{-1}, it is not enough to lift the parcels to their LFC. Due to the absence of large-scale forcing (Fig.8.2) or enough low-level convergence, one may conclude that deep convection is not likely without an additional mechanism that removes the existing low-level inversions or lids. Thus, despite the high conditional instability observed in the area, the high inhibition does not allow the initiation of deep convection in this day.

8.1.5 Summary and conclusions

The main purpose of this analysis is to try to answer the question of why deep convection did not develop in the area despite the existing high conditional instability in the CSIP domain. Although a wet near-surface was observed in the CSIP area - with respect to the distribution in the UK - low temperatures at the near-surface and a dry boundary layer did not favour deep convection. Additionally, the boundary layer remained capped by a strong lid whose space-time evolution was described by serial sonde ascents from Chilbolton, Reading, Swanage, and Preston Farm. The little warming observed in the PBL could not resolve the existing low-level inversion which remained mostly unchanged in the CSIP area for the period investigated. The absence of large-scale forcing and/or sufficient low-level convergence did not allow parcels to break through the low-level lid. One may conclude that in this event the inhibition is too high although CAPE is available and near-surface convergence is observed.

It is hypothesized that the deep convection observed in the western region of the UK is due to advection of very moist air from the south-west, which results in a situation of very high conditional and potential instability with CAPE over 2000 J kg^{-1}, LI < -6 °C and KO $\sim$ -8 K, and low inhibition, in addition to the presence of a frontal convergence zone which triggered convection. On this day, the IWV and boundary layer moisture in the CSIP domain is comparable to or even higher than the observations on the IOP5 (Figs.6.4 and 6.7) in areas where deep convection was observed. However, on the IOP2, high inhibition is observed in the CSIP domain whereas on the IOP5 CIN $>$-15 J kg^{-1} was shown in the whole area at the time when initiation took place. This highlights the importance of evaluating the atmospheric conditions, such as stability, available potential energy and inhibition, rather than the temperature or moisture distributions. Moreover, the key role of the lids in the inhibition of deep convection is pointed out in this analysis showing that the existence of high CAPE and/or PBL convergence do not assure the occurence of deep convection.

8.2 CSIP IOP8: Sea breeze case

It is well known that induced circulations such as sea breezes frequently affect convective initiation in the UK (Bennett et al., 2006). However, little is known about sea breeze convection in the UK. The UK receives relatively low insolation and has relatively strong geostrophic winds compared with the above-mentioned locations. For these reasons, sea breeze convection occurs less frequently and is generally not as intense as observed elsewhere. For the UK, it is unclear how significant this mechanism is relative to other modes of convection initiation (Bennett et al., 2006).

During CSIP IOP8, July 13, 2005, the cloud-free conditions and high insolation allowed the development of sea breezes from all coasts. The T+15 forecast from the 12 km Met Office mesoscale model shows a convergence line extending from Exeter to London, and beyond to The Wash, at 1500 UTC. The band of surface convergence associated with the sea breeze penetration inland is also a well-defined feature in the NIMROD-10 meters wind analysis for 1700 UTC (Fig.8.9). NIMROD is a fully automated system for weather analysis and nowcasting based around a network of C-band rainfall radars (Golding, 1998). It is interesting to study whether the resolution of the CSIP observations is enough to obtain a good representation of the convergence lines as the sea breezes penetrate inland, and to verify if the position and strenght of the convergence lines in the 12 km Met Office mesoscale model are consistent with the surface observations. Focus will be on the following issues,

1. The effect of the sea breeze on convection, and initiation or enhancement along the front.

2. The influence of moisture and moisture transport on the development of convection.

3. The influence of the sea breeze on the atmospheric stratification and lid evolution.

4. Whether subtle differences in temperature and moisture can lead to convection breaking through lids in some places but not in others.

5. The influence of the sea breeze on the convection indices.

6. The applicability of the combination of convection-related indices (chapter 7) to predict the areas where deep convection is observed.

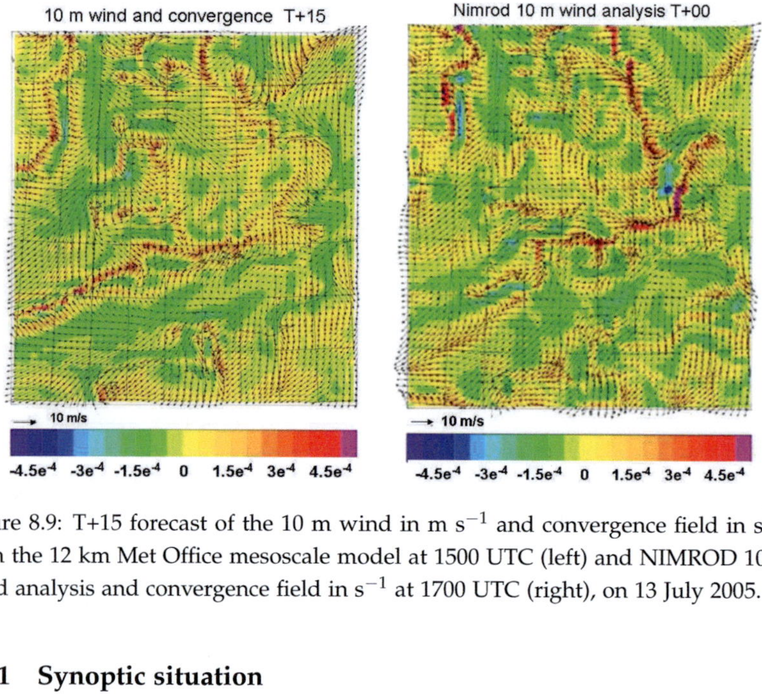

Figure 8.9: T+15 forecast of the 10 m wind in m s^{-1} and convergence field in s^{-1} from the 12 km Met Office mesoscale model at 1500 UTC (left) and NIMROD 10m wind analysis and convergence field in s^{-1} at 1700 UTC (right), on 13 July 2005.

8.2.1 Synoptic situation

The surface analysis for 1200 UTC shows a ridge extending across the central UK and a cold front advancing slowly from the north (Fig.8.10). The MSG infrared (IR) imagery (10.5-12.5 ţm, representing the cloud top temperature) shows a band of high clouds which extends along a line oriented W to E in central UK (Fig.8.11). This band of high clouds corresponds to the cirrus canopy associated with the advance of the cold front from the north. A very weak north-north-westerly flow covered the CSIP area and surface temperatures reach 29 °C in some of the CSIP radiosonde stations. The 500 hPa omega field at 1200 UTC shows that large-scale-forcing is of minor importance in the CSIP domain but lifting of 0 to -8 hPa h^{-1} concentrated in a south-west north-east line across central UK is observed. The maximum synoptic lifting of -8 hPa h^{-1} occurs in the north-eastern area of that line.

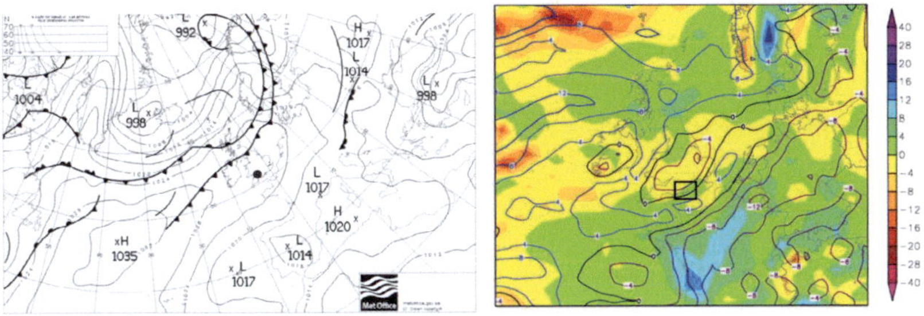

Figure 8.10: Met Office analysis of sea level pressure (left) and GFS analysis of omega field in hPa h^{-1} at 500 hPa (colour scale) and KO-index in K (isolines) (right) at 1200 UTC on 13 July 2005.

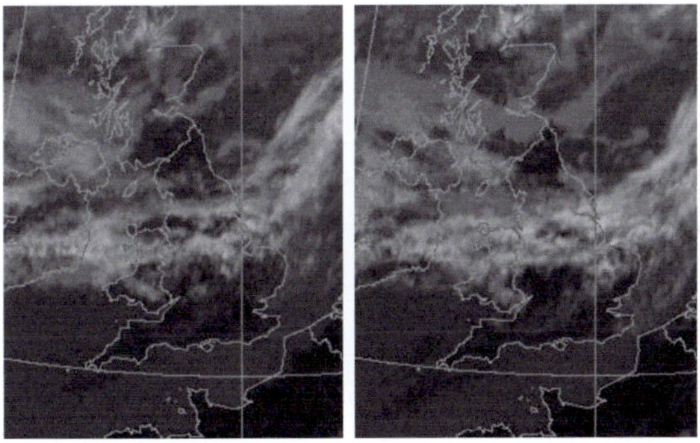

Figure 8.11: High-resolution MSG IR imagery at 1200 UTC and 1400 UTC on 13 July 2005.

8.2.2 Inland penetration of the sea breeze: detection by surface and PBL observations

Near-surface measurements and radiosoundings are used to investigate whether the resolution of the CSIP observations is high enough to adequately detect the inland advance of sea breezes and whether the main temperature and moisture variations at the low-levels are related to them.

Near-surface observations

The use of SYNOP stations and AWSs with 2 m AGL measurements build in a very dense network of 25 stations around Chilbolton, allows the analysis of the surface temporal evolution and the spatial distribution of temperature, humidity, and wind field. Fig.8.12 shows the near-surface temperature and moisture evolution in the period from 0800 to 1600 UTC.

At 0800 UTC, a temperature gradient from south to north is observed. Lower temperatures of about 18 °C in the south contrast to around 20 °C elsewhere. By this time, mixing ratio values vary between 9.5 g kg^{-1} in the south-east to 11.5 g kg^{-1} at the north of Chilbolton and Larkhill (Fig.8.12). The wind field shows highly variable winds (Fig.8.13). In the north-eastern region, a SW-NE oriented convergence zone passing over Larkhill is formed. The sea breeze is not a defined feature in the observations, and the convergence line is not related with any sea breezes.
By 1000 UTC, temperature has increased in the whole area. Especially, in the area of Larkhill, Chilbolton, and Reading an increase of around 5 °C is observed. A temperature difference of about 5-7 °C is established between the central area, Chilbolton, Larkhill, and the eastern and southern coasts (Fig.8.12). The temperature gradient favours the formation of secondary circulation systems. Strong southwesterly winds from the southern coast line of about 3.5 m s^{-1} indicate the position of the sea breeze penetrating inland from that coast (Fig.8.13). In the north-east, above 51.6° N, wind has turned from south-west to south-east probably corresponding to the development of the sea breeze at the eastern coast. At about 0° longitude, a convergence line with N-S orientation is formed.
By 1200 UTC, temperature in the central and northern areas has increased to around 27 °C, while in the south the sea breeze penetration causes a decrease of temperature and an increase in moisture. The sea breeze from the southern coast is still far from Chilbolton and Larkhill, however, Larkhill shows the wettest near-surface, 13 g kg^{-1}, whereas at Chilbolton it is 2 g kg^{-1} drier.
By 1300 UTC, at about 52° N, the N-S oriented convergence line observed initially at 1000 UTC, moves to the west because of the inland penetration of the east coast sea breeze. Still, no changes are observed in the temperature and moisture field in the area, but convergence increases to about -0.35 x 10^{-3} s^{-1}. The convergence in this area is stronger than the one associated with the sea breeze penetration from the south.

By 1400 UTC, a temperature difference of nearly 10 °C exists between the southern areas and inland. In the southern region the sea breeze inland penetration has reduced the temperature to approximately 19 °C, whereas in some areas in the north nearly 29 °C are measured.
By 1600 UTC, the sea breeze has passed Chilbolton accompanied by an increase in moisture and a decrease of temperature at this station. The inland penetration of the east coast sea breeze causes a temperature decrease of about 2 °C and a moisture increase of approximately 0.5 - 1 g kg^{-1}.
One may conclude that the near-surface temperature and moisture distributions are mainly dominated by the inland penetration of the sea breezes from the southern and eastern coasts. At the AWS-2 meters wind analysis (Fig.8.13), the location and strength of the convergence lines associated with the s ea breezes' from the southern and eastern coasts is consistent with the NIMROD-10 meters wind analysis and the 12 km Met Office mesoscale model (Fig.8.9).

Boundary layer observations

Detailed observations of the conditions in the boundary layer are limited to the area where radiosondes are available (Fig.8.14). As representative of the lower boundary layer, the temperature and moisture mean of the lower 50 hPa is shown. The spatial temperature distribution results from the interpolation of radiosondes in the area, and the mixing ratio distribution from the combination of radiosonde and GPS measurements. Spatial distributions are shown at 1000, 1200 and 1400 UTC when radiosondes are available at all stations.

The temperature field (Fig.8.14a) shows a continuous increase from south to north in agreement with the near-surface observations. Maximum temperatures of about 23-24 °C are measured in the areas of Larkhill, Chilbolton and Reading at 1400 UTC, whereas in the south and south-east temperatures remain below 20.5 °C for the whole period. Due to the low resolution of the radiosonde network in comparison to the surface measurements, the inland penetration of the sea breeze from the southern coast is not a well-represented feature at the temperature field. The maximum mixing ratio (Fig.8.14b) in the area is observed along a SW-NE oriented line passing through Larkhill, where at 1200 UTC a maximum humidity of about 12.4 g kg^{-1} occurs. A difference in the order of 1-2 g kg^{-1} is observed for the whole period between Larkhill and its closest radiosonde stations, Bath and Chilbolton, approximately 25-50 km away. Despite the inland penetration of the southern coast sea breeze, which is accompained by an increase in moisture, the lowest mixing ratio values, approximately 10-11 g kg^{-1} are observed for the whole

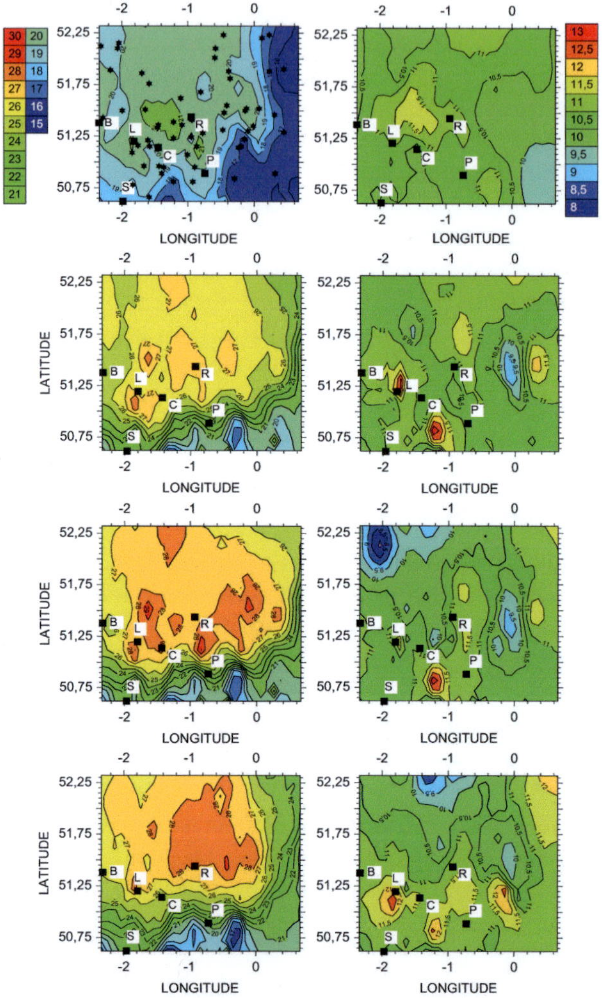

Figure 8.12: Spatio-temporal evolution of near-surface temperature in °C (left) and mixing ratio in g kg^{-1} (right) at 0800, 1200, 1400, and 1600 UTC (from top to bottom) on 13 July 2005. Black squares indicate the positions of the radiosonde stations. The black stars indicate the positions of the AWSs and the SYNOP stations.

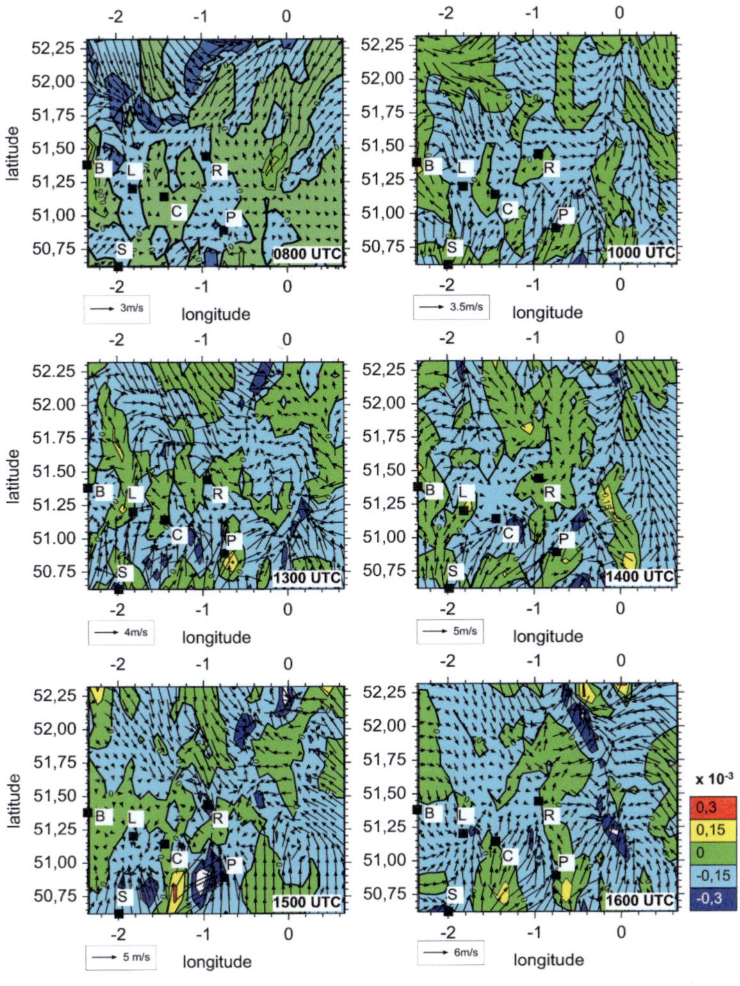

Figure 8.13: Near-surface horizontal wind divergence x 10^{-3} in s^{-1} and horizontal wind vectors in m s^{-1} on 13 July 2005. The positions of the radiosonde stations are indicated by black squares.

period in the area close the coast. From the analysis of the surface measurements (Figs.8.12, 8.13), the inland penetration of the sea breeze and consequent changes in the temperature, moisture and wind fields, can be clearly identified. However, the spatio-temporal distribution of PBL's moisture shows that the areas with higher

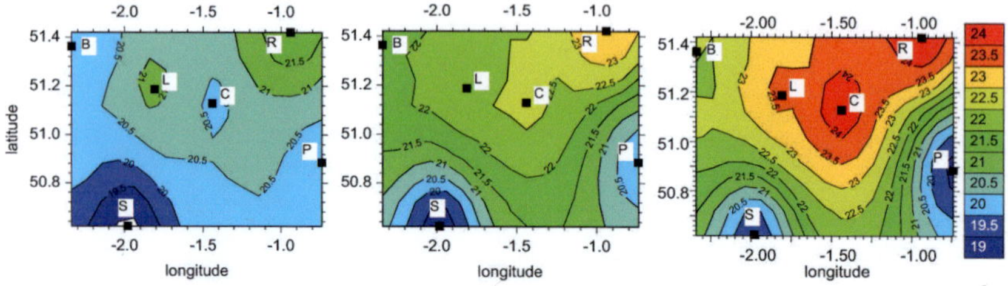

a)

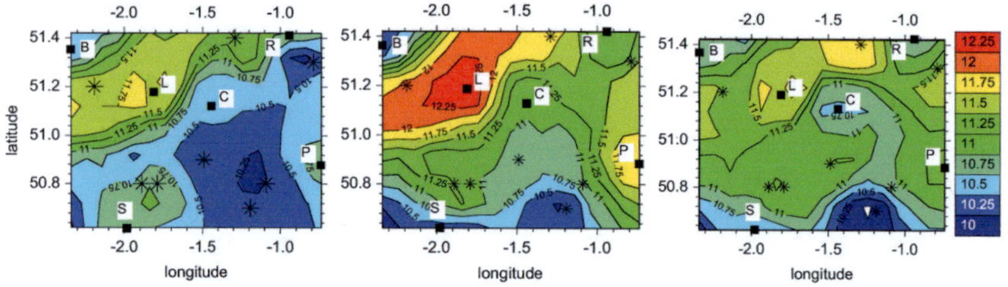

b)

Figure 8.14: Spatial distribution of the 50 hPa ML mean of the lower boundary layer a) temperature in °C (from interpolation of radiosonde data), and b) mixing ratio in g kg^{-1} (from combination of radiosonde and GPS data), at 1000 UTC (left), 1200 UTC (center), and 1400 UTC (right) for the CSIP domain on 13 July 2005. Black squares indicate the positions of the radiosonde stations. Stars indicate the positions of GPS stations.

water vapour content are not related with the sea breeze penetration inland. On this day, the IMK wind lidar was located at Chilbolton. Its RHI-scans (Fig.8.15) and PPI-scans (Fig.8.16), allow a more detailed analysis of the inland penetration of the sea breeze from the south towards Chilbolton. In Fig.8.15, RHI-scans are shown at 1508 UTC and 1530 UTC, in the period when the sea breeze, from the south, arrives at Chilbolton. The range of measurement is approximately 6 km. The aerosol backscatter measurements give an indication of the cloud base. At 1508 UTC, the sea breeze is observed 4 km south of Chilbolton. Above the sea breeze and ahead of it, northerly winds are measured. By 1530 UTC, the sea breeze has advanced reaching Chilbolton. The height of the front is between 800 and 1000 m AGL. The speed of progression is in the order of 3.5 m s^{-1} or 12.6 km h^{-1}. The width of the sea breeze front is about 2 km. In front of the sea-breeze nose a cloud-base level is detected at about 3.5 km AGL. Near the surface, the AWS measurements at Chilbolton (Fig.8.17) show in the period from 1500 to 1600 UTC a temperature

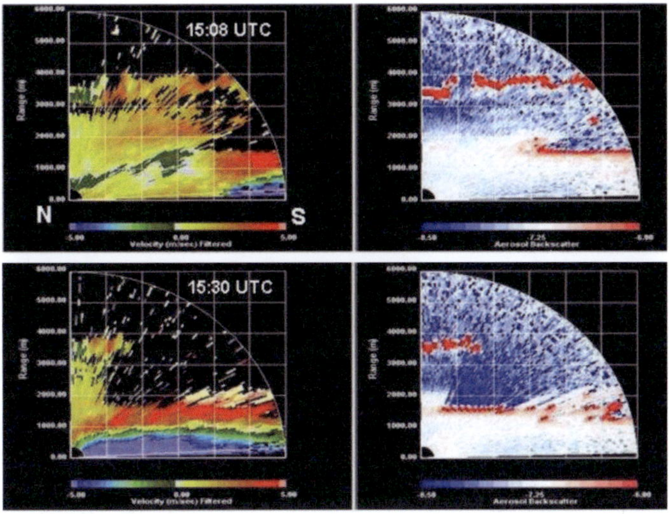

Figure 8.15: RHI-scan towards south, 180°, at 1508 UTC (top) and 1530 UTC (bottom) from the IMK wind lidar located at Chilbolton measuring horizontal wind (left) and aerosol backscatter (right). Negative wind speeds indicate winds directed towards the station.

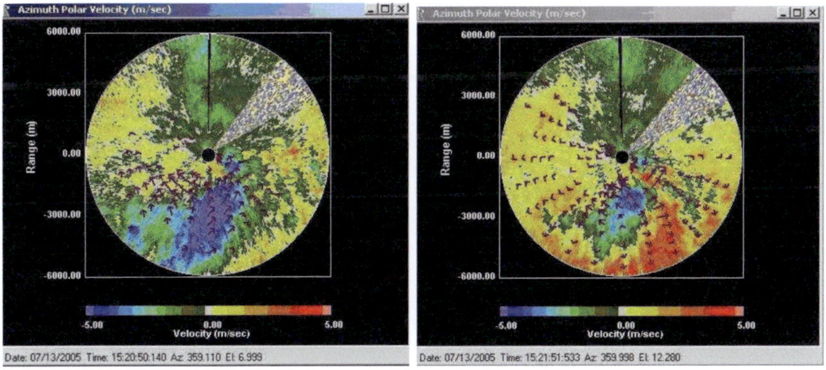

Figure 8.16: PPI-scan at 1520 UTC at 7° (left) and 13° (right) of elevation from the IMK wind lidar located at Chilbolton.

decrease from 28.2 °C to 25.5 °C. Dew point temperature increases from 13.9 °C to 16.5 °C. The wind turns from north-westerly (310°) to south-westerly (245°) direction and the wind speed increases from 1 m s^{-1} to 4 m s^{-1}. In agreement with

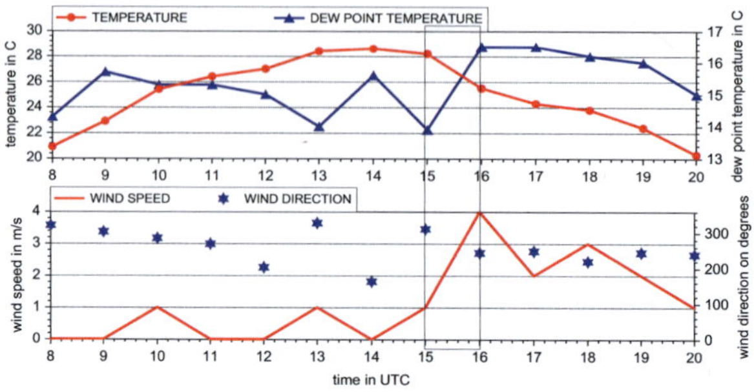

Figure 8.17: Temporal evolution of near-surface wind speed (m s^{-1}), wind direction (°), temperature, and dew point temperature (°C) from AWS measurements at Chilbolton for the period from 0800 to 2000 UTC on 13 July 2005.

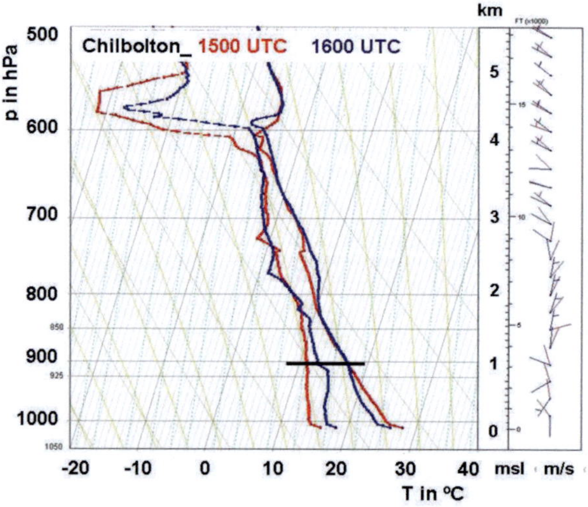

Figure 8.18: Arriving of the sea breeze at Chilbolton radiosonde station on 13 July 2005, as seen from Chilbolton radiosonde profiles at 1500 (red) and 1600 UTC (blue). The full black horizontal line indicates the depth of the sea breeze. Wind barbs in m s^{-1} are shown on the right hand of the diagram (half-barb=2.5 m s^{-1}; whole-barb=5 m s^{-1}).

the RHI-scans (Fig.8.15), the radiosonde profiles at 1500 and 1600 UTC (Fig.8.18) show that the height of the sea breeze is approximately 900 m AGL. At 1600 UTC, one may observe a temperature decrease of about -1.5 K and a moisture increase of $\sim$ 1.0 g kg^{-1} in this layer. The PPI-scan (Fig.8.16), at 1520 UTC at an elevation of 7° shows that the inland penetration of the sea breeze is not uniform, i.e., the sea breeze front does not advance parallel to the coast line. South-westerly winds are observed in the south and south-west of Chilbolton, but north-easterly winds in the south-east. In Figs.8.12, 8.13, the uneven penetration of the sea breeze inland can also be observed. At an elevation of 13° only the sea breeze front is seen surrounded by northerly winds opposed to the sea breeze advance.

8.2.3 Spatio-temporal evolution of atmospheric water vapour in relation to the wind flow conditions

As we have seen in the last section, the inland penetration of the sea breezes has a significant impact on the water vapour distribution observed in the lower atmospheric levels. However, the area where the maximum water vapour content is observed is not related to the advance of the sea breezes inland. Since areas of high temperature and moisture favour the occurrence of deep convection, it is interesting to investigate the origin of the SW-NE oriented band of high PBL moisture (Fig.8.14b) observed in the investigation area using GPS IWV measurements and radiosondes.

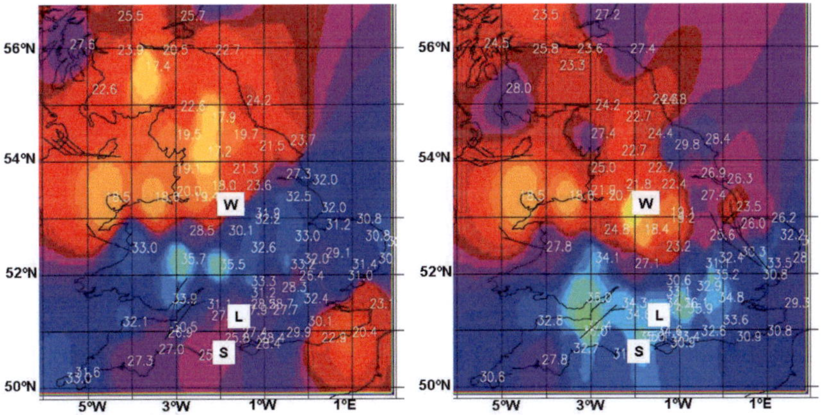

Figure 8.19: GPS-derived IWV in kg m^{-2} and 2 km wind at 0600 and 1600 UTC on 13 July 2005.

Ahead of the cold front (Fig.8.10), an area of high moisture is present in a SW-NE oriented line across central UK (Fig.8.19). The area of high moisture was present from the previous day. From 0600 to 1600 UTC, the north-north-westerly flow advects the moist air mass approximately 100 km southward. In Fig.8.20, 3 radiosonde profiles representing the 3 different air masses observed in the area at about 1200 UTC are compared. The following radiosondes are used for the comparison: Watnall representing the dry air mass behind the high moisture band, Larkhill representing the air mass of the band of high moisture advected from the north, and Swanage representing the southern areas close to the coast where the band of high moisture has not arrived yet but the sea breeze from the southern coast has passed over at about 1000 UTC. The positions of these radiosonde stations are indicated in Fig.8.19. At Watnall north-westerly winds can be seen from the near-surface to upper levels, at Larkhill and Swanage the north-westerly flow is limited to the layer from about 700 hPa to upper levels. From about 950 hPa to 700 hPa wind comes from the north-east. At that time, the sea breeze from the south coast has already passed over Swanage but still has not reached Larkhill (Fig.8.13). The temperature profiles show no great differences in the three areas.

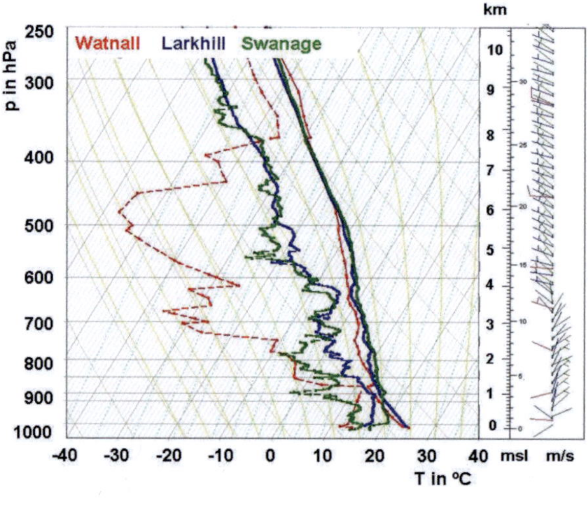

Figure 8.20: Comparison between radiosonde profiles at Watnall at 1115 UTC (red), Larkhill at 1200 UTC (blue), and Swanage at 1200 UTC (green) on 13 July 2005. Wind bars in m s^{-1} (half-barb=2.5 m s^{-1}; whole-barb=5 m s^{-1}).

Swanage shows lower temperature below 900 hPa coinciding with the depth of the sea breeze in the area (Fig.8.15). However, the water vapour content in the three radiosonde stations differs strongly. In agreement with the distinct wind regimes observed, the water vapour content of the atmosphere has been calculated from these radiosondes for the total column and for different atmospheric layers: from

300 to 700 hPa, where north-westerly winds are present in the three radiosonde stations, from 700 to 900 hPa, from 900 hPa to the near-surface, and from 950 hPa to the near-surface level. The results are summarized in Tab.8.3. A difference of almost 16 kg m^{-2} in IWV is found between Watnall and Larkhill and nearly 7 kg m^{-2} between Larkhill and Swanage. This strong difference in water vapour content occurs below 500 hPa. Despite the increase in moisture at Swanage below 900 hPa corresponding to the sea breeze depth (Fig.8.15), the water vapour content in Larkhill in the same layer is about 2 kg m^{-2} higher, in agreement with the near-surface (Fig.8.12) and PBL observations (Fig.8.14).

STATION	IWV total column	IWV 300-500 hPa	IWV 500-700 hPa	IWV 700-900 hPa	IWV 900hPa -surf	%	IWV 950hPa -surf	%
Watnall	18.9	0.5	0.7	7.8	9.8	51.7	5.9	31.5
Larkhill	34.6	1.3	6.2	13.6	13.4	38.6	7.4	21.3
Swanage	28.0	1.3	5.4	9.8	11.4	40.7	7.4	26.5

Table 8.3: IWV in kg m^{-2} for different layers, based on the radiosonde humidity profiles at Watnall (1115 UTC), Larkhill (1200 UTC), and Swanage (1200 UTC) on 13 July 2005.

The IWV measurements from GPS together with the IWV calculated from the available radiosondes are interpolated for the area of investigation on this day, 50.62° N to 52.32° N and 2.34° W to 0.66° E (Fig.8.21).

Early in the morning, by 0800 UTC (Fig.8.21a), the highest column-integrated water vapour, approximately 35 kg m^{-2}, is observed in the north-west, at about 52° N 2° W . At that time, there is a strong NW to SE horizontal gradient and IWV is in the order of 25 kg m^{-2} in the south-east. Two hours later (Fig.8.21b), the band of high moisture has been advected southwards and shows a SW to NE orientation. Larkhill and Reading are in the band of high moisture with values of about 33 and 32 kg m^{-2}, respectively. Chilbolton remains out of this area with an IWV value around 29 kg m^{-2}. By 1200 UTC (Fig.8.21c), the band of high moisture has moved more to the south. Close to the southern coast line, IWV has not changed significantly, despite the sea breeze penetration in the area. At 1400 UTC (Fig.8.21d), a difference of about 5 kg m^{-2} is found between Larkhill and Reading with IWV $\sim$ 35 kg m^{-2} and Chilbolton with 30 kg m^{-2}, which are between 25 and 50 km, respectively, away from each other. In the north-east, at about 52° N, the maximum IWV of the day is observed, around 36 kg m^{-2}. From that time on, the position of the high moisture band and the values measured do not change significantly.

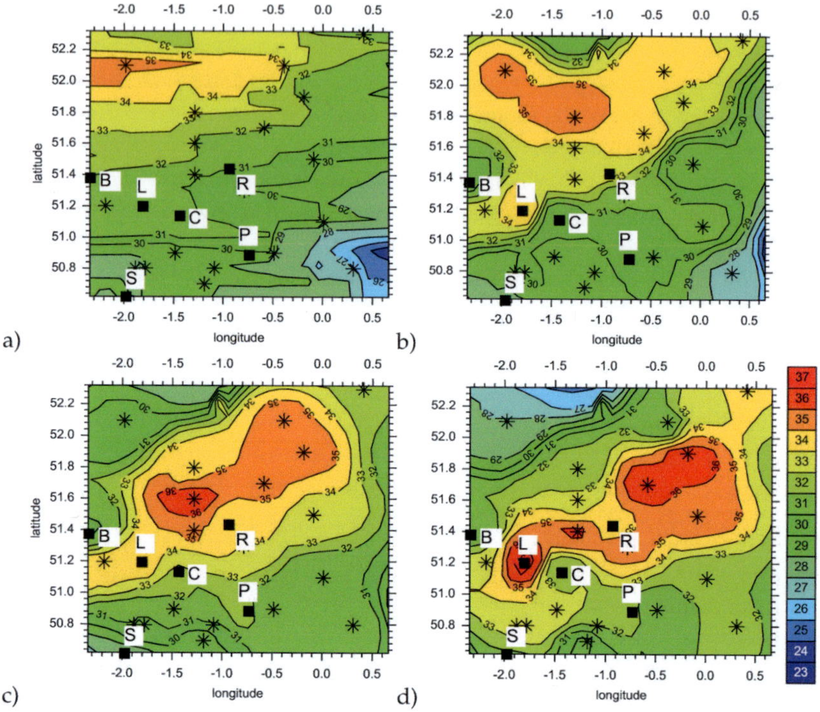

Figure 8.21: Spatially interpolated fields of IWV in kg m^{-2} from GPS and radiosondes in the area at a) 0800 UTC, b) 1000 UTC, c) 1200 UTC, and d) 1400 UTC on 13 July 2005. GPS stations are indicated by black stars, and radiosonde stations in the CSIP area are indicated by black boxes.

To clarify in which atmospheric layers the main temporal water vapour variations are observed, the vertical distribution of water vapour from radiosondes is investigated at Larkhill and Chilbolton (Tab.8.4). Larkhill is representative of the area where the band of high moisture is observed. The times for comparison are given by the availability of radiosondes. Vertical wind shear on that day is an important factor because the wind direction strongly influences the water vapour variations in the atmospheric layers through advection of dry/moist air. The wind flow regimes observed in each layer are summarized in Tab.8.4.

On that day, turbulent fluxes are not available to estimate the influence of evapotranspiration on the water vapour variation in the boundary layer.

The main observations are:

1. The main variations are observed below 500 hPa. Between 30 % and 40 % of the total-column water vapour is below the 900 hPa level and in most of the cases more than 20 % is below the 950 hPa level. Thus, the atmospheric water vapour content variability is dominated by variations in the boundary layer. These findings are in agreement with previous observations in this thesis.

2. The areas where higher moisture is observed are a consequence of advection of moist air from the north-east and are not related with the sea breeze advance inland from the southern coast.

3. At Chilbolton, the penetration of the sea breeze in the period from 1500 UTC to 1600 UTC results in an increase of $\sim$ 1.5 kg m^{-2} below 900 hPa.

4. Despite the increase in water vapour observed at Chilbolton at 1600 UTC, the total-column IWV at this time is still $\sim$ 1 kg m^{-2} lower than at Larkhill at 1400 UTC.

8.2.4 Lid evolution

A detailed analysis of the radiosonde profiles in the area shows the presence of several PBL and mid-tropospheric capping inversions. An interesting feature of this IOP is the variation in time and space of the various stable lids. Moreover, it is interesting to investigate whether the sea breeze influences the lid evolution and the effect of the lids in the inhibition of deep convection.

Early in the morning, the radiosonde profiles in the CSIP area show that multiple lids in the vertical are present, whose strengths vary significantly in time and space. As an example, the evolution of lids is shown at Chilbolton for the period from 1000 to 1600 UTC (Fig.8.22). The pressure level at the base of the lids and their strengths are shown in Tab.8.5. At 1000 UTC, a very strong lid of $\sim$ 3.0 °C, at 886 hPa and a weaker mid-tropospheric lid at about 580 hPa are present in the area (Tab.8.5). By 1200 UTC, an increase in both temperature and moisture below 900 hPa effectively erode the 886 hPa lid, but an inversion at 822 hPa is observed. In the period from 1400 to 1500 UTC, an increase in temperature below approximately 800 hPa removes the low-level lid, but there is a lid at about 600 hPa. From 1500 to 1600 UTC, the sea breeze has arrived at that station (Fig.8.15), reducing temperature and increasing moisture below 900 hPa (Fig.8.22, Tab.8.5). Despite the increase in boundary layer moisture, the mid-tropospheric lid, at about $\sim$ 580 hPa still constrains the evolution of deep convection. The evolution of the multiple lids observed at Chilbolton is similar to that at the other radiosonde sites in the investigation area with lids in the layer from 600 to 500 hPa. This has a significant influence on the evolution of cumulus clouds in the area.

STATION Larkhill	IWV total column	IWV 300-500 hPa	IWV 500-700 hPa	IWV 700-900 hPa	IWV 900hPa -surf	%	IWV 950hPa -surf	%
1000UTC	34.9	1.4	6.1	14.4	12.9	37.0	7.0	20.1
wdirection		NW	NW	NE	W		SW	
1200UTC	34.6	1.3	6.2	13.6	13.4	38.6	7.4	21.3
wdirection		NW	NW	NE	SW		SW	
1400UTC	37.2	1.4	5.5	17.8	12.5	33.5	6.9	18.6
wdirection		NW	NW	NE	NE		NE	

STATION Chilbolton	IWV total column	IWV 300-500 hPa	IWV 500-700 hPa	IWV 700-900 hPa	IWV 900hPa -surf	%	IWV 950hPa -surf	%
1000UTC	30.2	1.3	5.6	11.6	11.5	38.3	6.9	22.8
wdirection		NW	NW	NE	NE		NW	
1200UTC	32.9	1.4	5.8	13.4	12.3	37.3	7.2	21.9
wdirection		NW	NW	NE	NE		NW	
1400UTC	32.0	1.3	5.4	13.9	11.3	35.3	6.5	20.4
wdirection		NW	NW	NE	SE		SE	
1500UTC	33.4	1.3	5.5	15.1	11.5	34.5	6.6	19.7
wdirection		NW	NW	NE	NW		NW	
1600UTC	35.8	1.2	6.2	15.1	13.1	36.7	7.4	20.8
wdirection		NW	NW	NE	W		SW	

Table 8.4: IWV calculations in kg m^{-2} for different layers at Larkhill and Chilbolton radiosonde stations on 13 July 2005. Wind direction in each layer is included.

8.2.5 Combination of indices and the initiation of convection

Convection-related indices are calculated from the high-resolution moisture fields obtained by the combination of radiosondes and the IWV from GPS, lifting a parcel with characteristics of the 50 hPa mean of the lower PBL.

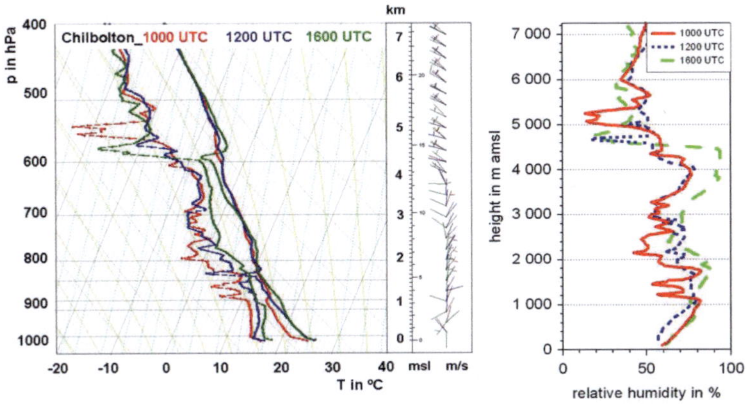

Figure 8.22: Skew T-log p diagram showing profiles of temperature and dew point temperature in °C at Chilbolton on 13 July 2005. Wind barbs in m s^{-1} (half-barb=2.5 m s^{-1}; whole-barb=5 m s^{-1}), and relative humidity in % (right) at the same time of the radiosonde profiles.

Station	Time	Pressure level in hPa	Lid strength in °C
Chilbolton	1000 UTC	886	3.1
		580	1.2
	1200 UTC	822	0.9
	1400 UTC	810	1.1
		513	0.9
	1600 UTC	580	0.2

Table 8.5: Pressure level at the maximum temperature at the inversion and capping strength at Chilbolton on 13 July 2005.

This combination at 1000 UTC (Fig.8.23) shows that in most parts of the investigation area the atmosphere is quite stable, as indicated by CAPE values close to zero and positive LI and KO-index, with absolute values of CIN $\geq$ 40 J kg^{-1} and LSI in the order of -1 to -2 °C in most parts. However, in the north-west in a W-E oriented band, low CAPE of 100 to 400 J kg^{-1}, LI of 0 to -1 °C, and KO index of -1.5 to -2 K, indicate low conditional and potential instability in this region. The maximum instability, CAPE $\sim$ 400 J kg^{-1} and lower CIN of about -20 J kg^{-1} are found in the western part of this band where high near-surface temperature and moisture - relative to the observations in the area - are seen (Fig.8.12). Vertical updraughts in the order of $w \geq 6$ m s^{-1} are needed to overcome the -20 J kg^{-1} CIN value. In this area, CAP is in the order of 1 to 1.5 °C and LSI is about 0 °C, showing that in the absence of strong large-scale forcing or PBL-convergence deep convection could be expected. The triggering temperature at this time has not been reached, Tc-Tv > 0 °C, and no large-scale lifting existed, thus a PBL-driven trigger mechanism is needed for parcels to be lifted to their LFC. The near-surface horizontal divergence field at 0800 UTC shows a convergence line extending from Larkhill to the north-east and stronger convergence in a west-east oriented line at about 51.8° N coinciding with the area where precipitation is observed at 1115 UTC.

By 1200 UTC (Fig.8.24), as a consequence of the PBL's temperature and moisture distribution (Fig.8.14b), a NE-SW oriented band is observed with higher CAPE of approximately 300-600 J kg^{-1}, LI ~ -1 to -2 °C, and KO-index ~ -1 to -3 K, and CIN, less than 20 J kg^{-1} and CAP ~ 1 °C, both lower than in the rest of the area. Despite the PBL's temperature and moisture increase in the period fron 1000 to 1200 UTC (Figs.8.12, 8.14), Tc is only reached in the area of Chilbolton. The MSG visible imagery at 1300 UTC shows that the band of cumulus has broadened at this time including the area of Chilbolton and extending towards the NE. The cloud band does not indicate the position of sea breeze front at the surface (Fig.8.12) but the position of the SW-NE oriented convergence line initially observed at 0800 UTC where at 1200 UTC higher CAPE and lower CIN than in the rest of the area are observed. Still, no heavy precipitation is observed along this band. Already at 1000 UTC, the near-surface horizontal divergence field (Fig.8.24) shows a N-S oriented convergence line at about 0.5° W in association with the east-coast sea breeze inland penetration.

At 1300 UTC (Fig.8.25), stronger convergence of ~ -0.3 x 10^{-3} s^{-1} is present at the north of this line, at about 52° N. In this area, large-scale lifting of about -4 to -8 h^{-1} and KO-index ~ -3 K are observed. The convergence zone coincides with the northern part of the band of relative moderate to high CAPE and low CIN - with respect to the observations in the area - and high low- and mid-tropospheric humidity (Fig.8.21).

At 1500 UTC, the MSG visible imagery and the radar rain rate show that some of the convective clouds were deeper by that time along this band producing some light precipitation. But the only significant showers are observed to the east of the area where surface convergence coinciding with CAPE values in the order of 500 to 600 J kg^{-1} and large-scale lifting of about -8 hPa h^{-1} were observed during our analysis. Probably, these showers initiated in the area where the surface convergence line and higher CAPE values were observed and the clouds were advected to the east in an area of high moisture (Fig.8.21) which favours the occurrence of heavier showers.

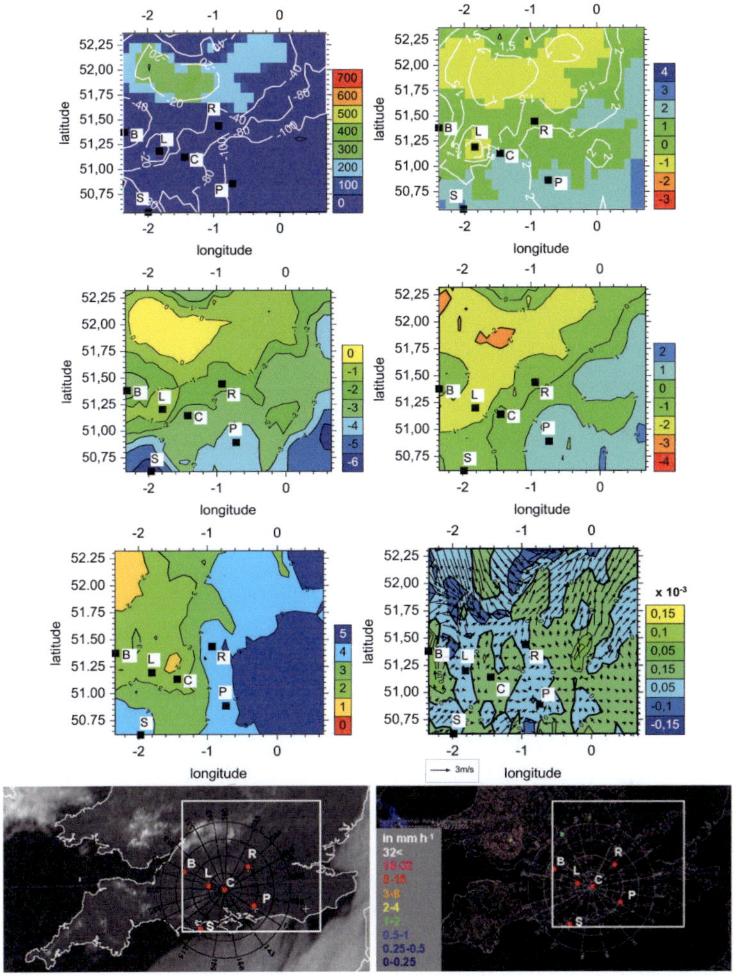

Figure 8.23: Combination of convection-related indices at 1000 UTC, satellite and radar observations at 1115 UTC on 13 July 2005. From top to bottom: CAPE and CIN in J kg^{-1} (left), LI and CAP in °C (right), LSI in °C (left), KO-index in K (right), Tc-Tv in °C (left), near-surface horizontal wind divergence x 10^{-3} in s^{-1} at 0800 UTC (right), satellite VIS imagery (left), and rain radar (right) at 1115 UTC. In the satellite and radar observations, the positions of the radiosondes are indicated by red dots, and the area of investigation is delimited by a white square. In the rest of the figures, the radiosonde stations are indicated by black squares.

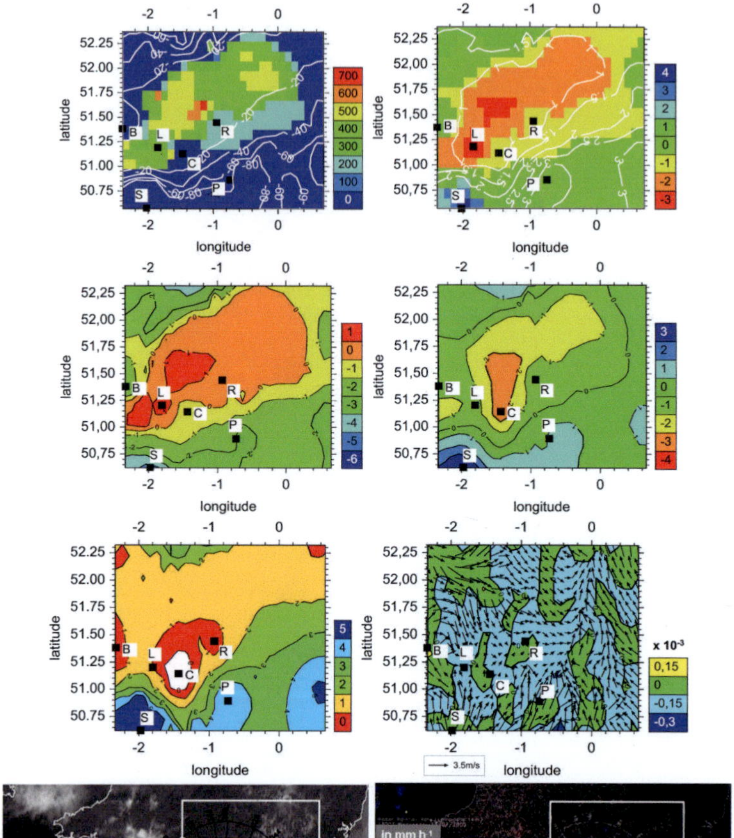

Figure 8.24: Combination of convection-related indices at 1200 UTC, satellite and radar observations at 1300 UTC on 13 July 2005. From top to bottom: CAPE and CIN in J kg^{-1} (left), LI and CAP in °C (right), LSI in °C (left), KO-index in K (right), Tc-Tv in °C (left), near-surface horizontal wind convergence x 10^{-3} in s^{-1} at 1000 UTC (right), satellite VIS imagery (left) and rain radar (right) at 1300 UTC. In the satellite and radar observations, the positions of the radiosondes are indicated by red dots, and the area of investigation is delimited by a white square. In the rest of the figures, the radiosonde stations are indicated by black squares.

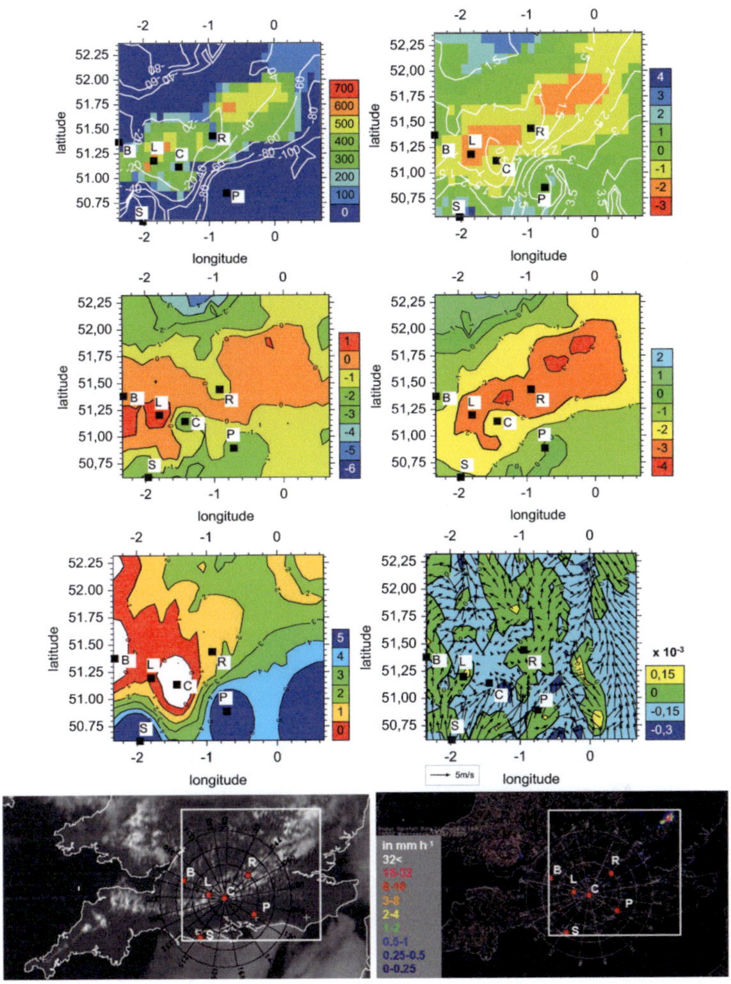

Figure 8.25: Combination of convection-related indices at 1400 UTC, satellite and radar observations at 1500 UTC on 13 July 2005. From top to bottom: CAPE and CIN in J kg^{-1} (left), LI and CAP in °C (right), LSI in °C (left), KO-index in K (right), Tc-Tv in °C (left), near-surface horizontal wind divergence x 10^{-3} in s^{-1} at 1300 (right), satellite VIS imagery (left) and rain radar (right) at 1500 UTC. In the satellite and radar observations, the positions of the radiosondes are indicated by red dots, and the area of investigation is delimited by a white square. In the rest of the figures, the radiosonde stations are indicated by black squares.

8.2.6 Summary and conclusions

On the day under discussion, convection developed over the CSIP area in association with the diurnal heating cycle. There were a lot of cumulus congestus clouds but only a few of them developed leading to brief light showers in the CSIP area. These occurred mainly in the north-east of the CSIP domain. Farther to the north-east there were a few heavier showers, particularly towards the Cambridge area.

Large temperature gradients between the coast and inland favoured the formation of sea breezes in the south and east coasts. Convection was affected by several rather weak warm lids, whose spatial and temporal variations in the area have been investigated. Variations in boundary layer temperature and moisture in the area were responsible for the eventual erosion of low-level lids, and the mid-tropospheric lids determined the extent of cloud growth. The analysis of the horizontal distribution of water vapour shows that cumulus formation occurred in the areas where high water vapour content and convergence lines superimpose.

Horizontal convergence played a decisive role in the initiation of convection on this day. Early in the morning, a surface convergence line with SW-NE orientation is the reason for the development of the cumulus clouds along this band. The convergence line associated with the southern coast sea-breeze, with an associated convergence of about -0.2 x 10^{-3} s^{-1}, enhanced convection where uplift occurred at the sea breeze head, however, this uplift was not enough to erode the existing mid-tropospheric lids. The convergence associated with the east coast sea breeze was stronger, approximately -0.35 x 10^{-3} s^{-1}. The showers that formed between 50 and 100 km north-east of Chilbolton may have been associated with convective clouds that were initiated along this surface convergence line, at about 52° N. Since radiosonde profiles are not available in this area, it is hypothetized that the strong surface convergence and the large-scale lifting in this area lead to the erosion of the mid-tropospheric lids producing heavier showers.

This leads to the assumption that the sea-breeze alone did not cause severe weather, and that additional driving forces are required to enhance the destabilization processes and to resolve existing lids. The combination of GPS and radiosonde measurements and the analysis of the combination of convection-related indices presented have been proven to be useful for the prediction of deep convection likelihood.

8.3 CSIP IOP16 and COPS IOP8b: locally initiated storms over flat and complex terrains

This section aims at identifying and comparing the processes that lead to the formation of isolated showers in the relatively flat CSIP area, and in the more complex terrain of the COPS experiment. Furthermore, the usefulness of the combination of GPS and radiosonde data is evaluated for a complex terrain.

Initially, the synoptic situation is described. The near-surface and PBL conditions are investigated to gain some knowledge about the spatial and temporal variations of temperature, moisture, and wind and their effect on the convection initiation in the area. The spatial and temporal evolution of the atmospheric water vapour is analyzed by means of IWV measurements from the GPS network. Finally, the spatial distribution of convection-related indices is calculated and their temporal evolution is investigated in relation with the temperature and moisture variations observed in the boundary layer.

The convection-related indices are combined to analyze deep convection likelihood in the area. IOP16 from the CSIP campaign and IOP8b from the COPS campaign have been selected as the appropriate days for this comparison. On both days, warm and mainly sunny conditions are observed, and isolated showers develop and intensify.

8.3.1 CSIP IOP16

Synoptic situation and convective activity

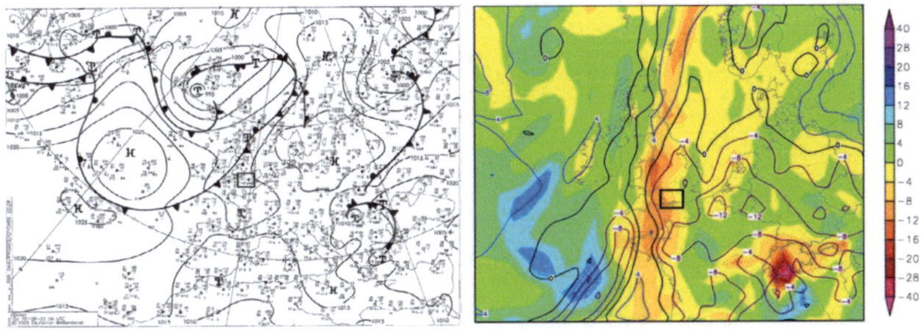

Figure 8.26: DWD analysis of mean of sea level pressure (left), and GFS-analysis of vertical motion in hPa h^{-1} at 500 hPa (colour scale) and KO-index in K (isolines) (right), at 1800 UTC on 18 August 2005.

During the IOP16, on the August 18, 2005, the CSIP area was covered by a weak southerly flow ahead of a cold-front marked by a cloud belt advancing slowly from the west. From 1200 UTC onwards, at the 500 hPa level the CSIP area suffers large-scale forcing with moderate lifting of -2 to -8 hPa h^{-1} concentrated in a north-south oriented line across UK (Fig.8.26). Much of the area experienced a warm, mainly sunny day with only shallow convection. The first cumulus, around 1100 UTC, were tied very closely to topography in south-east England and the south Midlands (Fig.8.27). The cumulus became soon more widespread. Around 1515 UTC, the early phase of the clouds that developed into isolated showers is observed to the west of Chilbolton. At this time, the main band of cold-frontal rain is still far from this area (Fig.8.27). At 1700 UTC, an area of significant showers can be seen on the radar network (Fig.8.27). These showers continued to intensify while still remaining isolated.

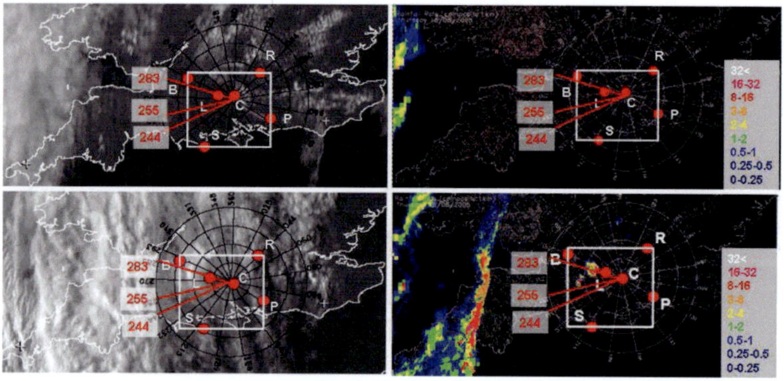

Figure 8.27: MSG satellite (left) and rain-radar (right) at 1200 (top) and 1700 UTC (bottom) on 18 August 2005. The positions of the radiosondes are indicated by red circles, and the area of investigation is delimited by a white box. Azimuths 244, 255 and 283 degrees are highlighted in red (referred to later in the text.)

GPS-derived spatio-temporal evolution of atmospheric water vapour content

The spatial and temporal evolution of IWV from GPS and radiosondes is presented in Fig.(8.28) for the area of investigation in the period from 0900 to 1700 UTC.

Early in the morning (Fig.8.28a), the distribution of IWV is quite homogeneous with values ranging from 21 to 23 kg m^{-2}. Three hours later (Fig.8.28b), the water vapour content has increased only by about 2 kg m^{-2} in the north-west. At 1400 and 1500 UTC (Fig.8.28c,d), a gradient of increasing IWV from south-east, around 23 kg m^{-2}, to north-west, 26-27 kg m^{-2} is observed. By 1600 and 1700 UTC (Fig.8.28e,f), the observed gradient is larger. Whereas IWV has not changed much in the south-eastern region, about 24-25 kg m^{-2}, in the north-western area, in the area between Larkhill and Bath, a maximum of about 32-33 kg m^{-2} is observed.

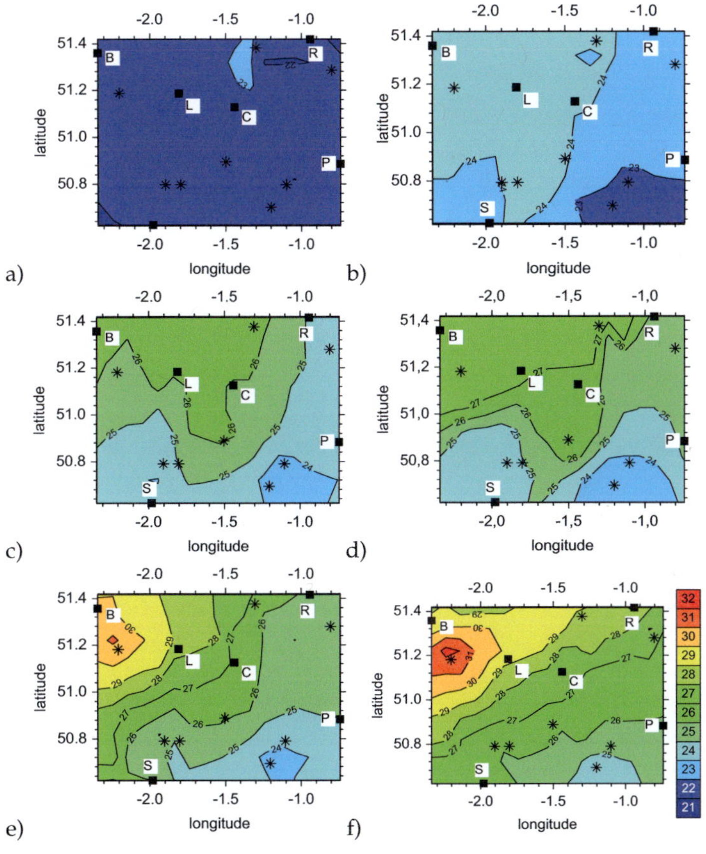

Figure 8.28: GPS-derived IWV in kg m^{-2} at a) 0900 UTC, b) 1200 UTC, c) 1400 UTC, d) 1500 UTC, e) 1600 UTC, and f) 1700 UTC on 18 August 2005.

The analysis of the IWV distribution in the UK and the wind flow in the area show that advection of moisture from the west-south-west is responsible for the increase in moisture observed in the investigation area (not shown).

Near-surface and boundary layer observations

In Fig.8.29, the spatial distribution of the near-surface temperature and mixing ratio in the period from 1200 to 1700 UTC is shown. For the interpolation, stations outside the area of investigation are used. The lack of surface stations in the north-western region does not allow the investigation of the small-scale moisture and temperature variability in this area.

No significant changes in the distribution of temperature and water vapour content at the near-surface are observed in the period under discussion. During the day, a temperature gradient of increasing temperatures from south to north exists. In the northern areas, temperatures are higher in the east. During the whole day, mixing ratio distribution at the near-surface shows maximum values in the southern areas, about 11.5 to 13 g kg^{-1}, decreasing towards north with minimum measurements around 8 g kg^{-1}. The near-surface observations in this case disagree with the IWV spatio-temporal evolution, the latter showing higher atmospheric water vapour content in the north-western areas decreasing towards the south-east.

Radiosonde profiles from the six radiosonde stations are only available at 1200 and 1700 UTC. Fig.8.30 shows the PBL temperature and mixing ratio distribution in the investigation area for these times. To obtain a representative PBL temperature and mixing ratio value, the 50 hPa mean of the lower boundary layer is calculated. For comparison, mixing ratio from the interpolation of radiosondes and mixing ratio after applying the adjustment method are shown. In general, the PBL's temperature spatial distribution (Fig.8.30a) agrees with the near-surface observations (Fig.8.29) showing higher temperatures in the eastern stations. By 1200 UTC, minimum temperature around 19.5 °C are in the area of Swanage, and higher temperatures of about 22.5 °C, are found in the areas of Reading and Preston. By 1700 UTC, the temperature has increased at most of the stations, especially at Reading, where an increase of about 2 °C is observed. However, in the area of Bath, a temperature decrease of approximately 1.5 °C occurs. This is probably due to the reduction in insolation as the frontal cirrus spreads in from the west. The water vapour content (Fig.8.30b) in the lower troposphere shows a different pattern than the temperature distribution, with wetter areas in the west and driest areas in the east. Although the Area of Deep Convection (ADC) was close to the radiosonde sites Bath and Larkhill, it did unfortunately remain between these sites for much of its early development. Therefore, the radiosondes at Larkhill and Bath did not sample the air mass of the ADC. However, the use of the GPS measurements to increase the resolution of the humidity field leads to interesting results. Both, at 1200 and espe-

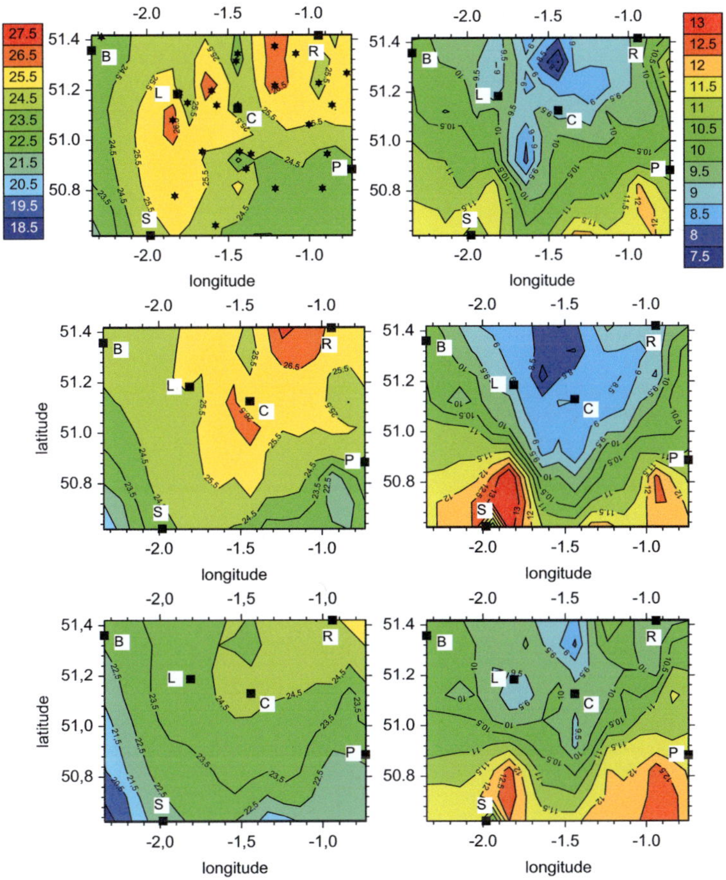

Figure 8.29: Spatio-temporal evolution of the near-surface temperature in °C (left) and mixing ratio in g kg^{-1} (right) at 1200, 1400, and 1700 UTC (from top to bottom), on 18 August 2005. The positions of the surface stations are indicated by black stars in the first diagram and the radiosonde stations are marked by black squares.

cially at 1700 UTC (Fig.8.30c), an area of high moisture is observed between both radiosondes in agreement with the total-column IWV distribution (Fig.8.28). At 1700 UTC, in this area the lower PBL mixing ratio is about 2-2.5 g kg^{-1} higher than at Bath and Larkhill. This strong increase in PBL water vapour content is due to advection of moisture in the boundary layer. Only when the resolution is increased, using in this case GPS measurements, an indication of the area where

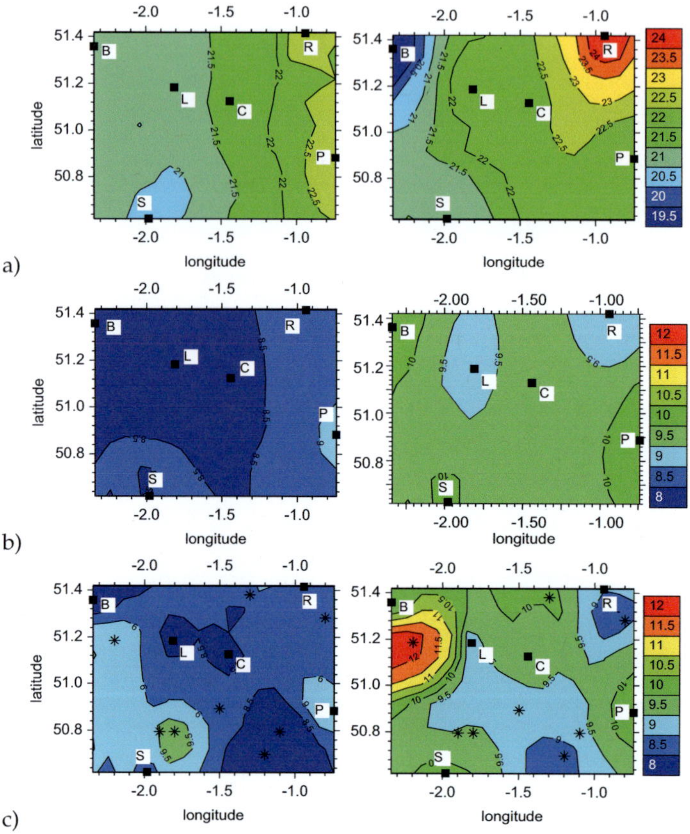

Figure 8.30: Lower boundary layer 50 hPa mean of a) temperature °C, b) mixing ratio in g kg^{-1} from interpolation of radiosonde data, and c) mixing ratio in g kg^{-1} from combination of radiosonde and GPS data, at 1200 (left) and 1700 UTC (right) on 18 August 2005. The positions of the radiosonde stations are indicated by black squares. The positions of the GPS stations are indicated by black stars.

deep convection is observed is obtained. Using the radiosonde profiles available at the stations of Bath, Larkhill and Preston Farm, height-time cross sections are interpolated (Fig.8.31). The evolution of potential temperature and mixing ratio at these stations is investigated for the period from 1100 UTC to 1700 UTC. Bath and Larkhill stations are close to the area where deep convection occurs. Preston Farm is located to the east of the investigation area where widespread clouds form in the morning but no precipitation is observed (Fig.8.27). The height-time cross-section

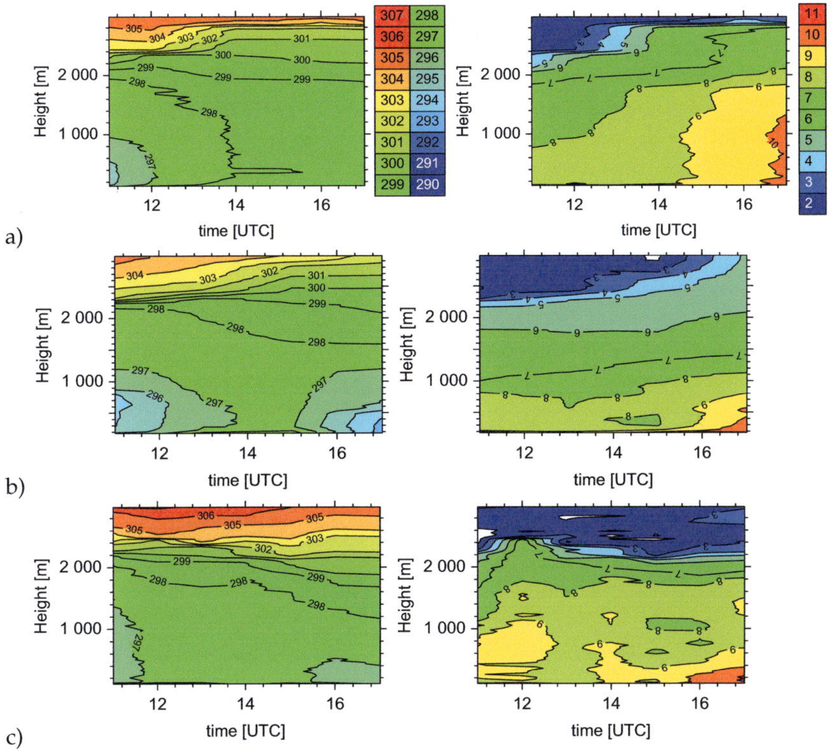

Figure 8.31: Height-time cross-sections of potential temperature in K (left) and mixing ratio in g kg^{-1} (right) at a) Bath, b) Larkhill, and c) Preston Farm on 18 August 2005.

at Bath (Fig.8.31a) shows a well-mixed boundary layer below approximately 1000 m amsl with almost constant values of 8 g kg^{-1} from 1100 to 1500 UTC. In this period, an increase in temperature in this layer takes place, from 295 K at 1100 UTC to 298 K at 1500 UTC, approximately. From this time, temperature decreases below 500 m amsl up to 294 K at 1700 UTC, but an increase of about 3 g kg^{-1} is observed in the lower boundary layer. At Larkhill (Fig.8.31b), a temperature increase in the PBL is also observed. Water vapour content in the PBL increases in time from about 8 g kg^{-1} at 1100 UTC, below 800 m amsl, to 10 g kg^{-1} at 1700 UTC, below 1800 m amsl, which is an indication of the convection activity in the area. At 1100 UTC, the PBL at Preston Farm (Fig.8.31c) is warmer than the ones at Bath and Larkhill, and temperature at that station remains almost constant through the day.

High mixing ratio values of about 9 g kg^{-1} are observed in the period from 1100 to 1200 UTC. At this time, mixing over the PBL can be seen up to 2500 m amsl. At 1400 UTC, vertical transport of moisture is observed up to 1500 m amsl. From that time, moisture in the layer from 500 m to 1500 m amsl decreases by about 1 g kg^{-1}. However, below the 500 m amsl level there is an increase of up to 11 g kg^{-1}.
From 1400 UTC, at Bath and Larkhill high values of mixing ratio and potential temperature are observed in the upper levels, from 1800 m to 2800 m amsl. This can be indicative of lifting in the area. Such variation is not observed at Preston Farm.

Horizontal winds and convergence

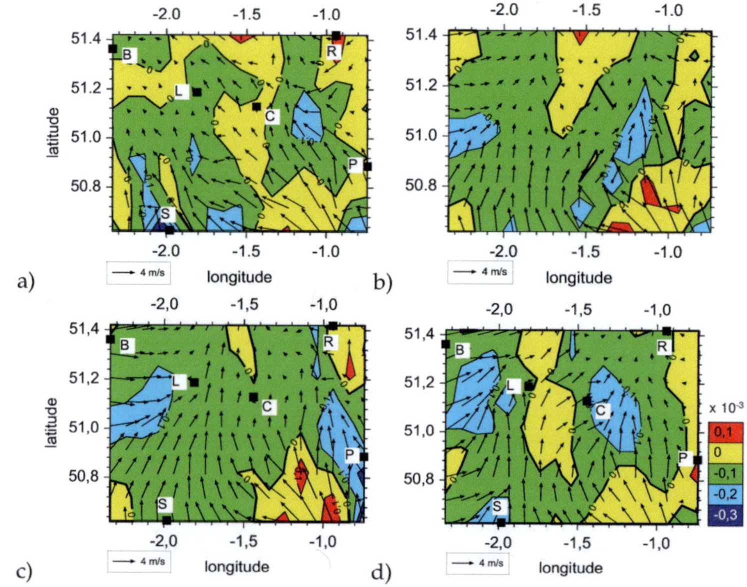

Figure 8.32: Near-surface horizontal wind divergence x 10^3 s^{-1} and wind vectors in m s^{-1} at a) 1200 UTC, b) 1400 UTC, c) 1500 UTC, and d) 1600 UTC on 18 August 2005.

The analysis of the near-surface wind field for the period from 1200 to 1600 UTC is shown in Fig.8.32. From 1400 UTC onwards (Fig.8.32b,c,d), an area of persistent surface horizontal convergence with orientation south-west north-east, and strength of about -0.2 x 10^3 s^{-1}, is observed between Bath and Larkhill radiosonde stations. The horizontal wind divergence field at the 950 hPa level in the PBL (not shown) at 1200 and 1700 UTC shows convergence in the eastern and western regions. By 1200 UTC, horizontal convergence is stronger in the eastern area whereas at 1700 UTC the convergence observed in the west is much stronger, -0.2 x 10^{-3} s^{-1}, and is localized in a NW-SE line between Bath and Larkhill coinciding with the area of high moisture previously observed (Figs.8.30, 8.28). Near-surface and PBL winds agree and show strong convergence between Bath and Larkhill prior to the development of deep convection.

Mid-tropospheric lids and their lifting by low-level and large-scale forcing

The evolution of the radiosonde profiles at Bath, in the west of the investigation area, and Reading, in the east of the investigation area is shown in Fig.8.33.

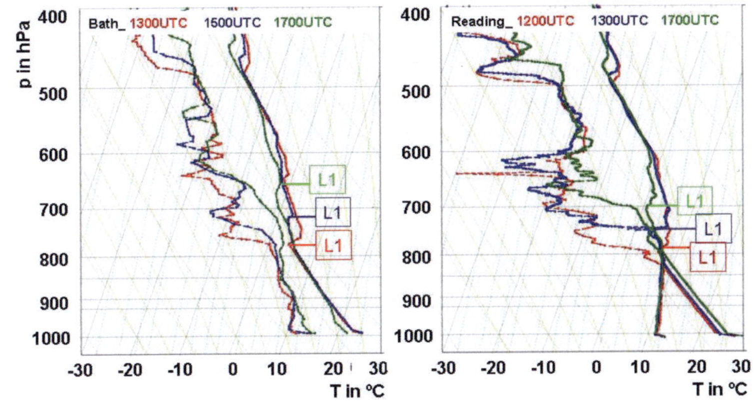

Figure 8.33: Skew T-log p diagram showing the radiosonde profiles at Bath (left), and Reading (right), on 18 August 2005. L1 indicates the position of the lid.

From early in the morning, mid-tropospheric lids at approximately 700 hPa/3 km and 500 hPa/6 km are observed at all of the radiosonde stations. By late afternoon, it can be derived from the radiosounding profiles (Fig.8.33) that a 500 m lifting in the whole troposphere is observed at both stations due to the combination of sufficient boundary-layer forcing (horizontal wind convergence in the area) and large-scale uplift. As a result of the lifting, the lid at 700 hPa/3 km weakens in the whole area. The path of a mixed parcel lifted from the lower 50 hPa is depicted in Fig.8.34 at Bath and Reading stations at 1700 UTC. At that time, the 3 km lid has been effectively eroded at Bath whereas at Reading the mid-tropospheric inversion is still present.

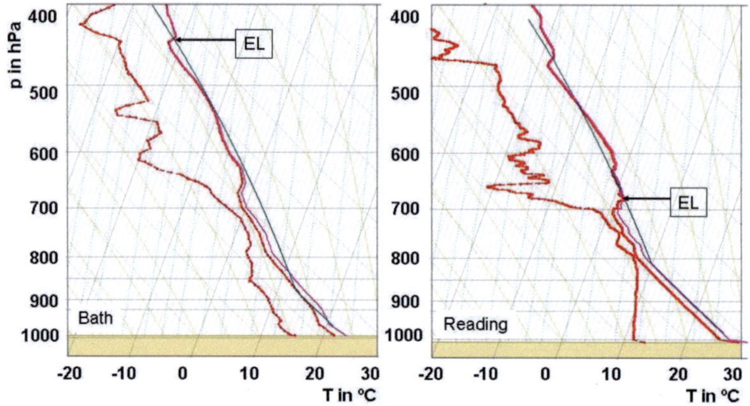

Figure 8.34: Skew T-log p diagram showing the radiosonde profiles at Bath (left) and Reading (right), at 1700 UTC on 18 August 2005. The pink line shows the virtual temperature, the green line indicates the path of a parcel lifted from the mean of the lowest 50 hPa mixed layer, and EL indicates the position of the equilibrium level.

Combination of convection-related indices

At 1200 UTC, the MSG visible imagery shows widespread cumulus but no deep convection in the eastern areas (Fig.8.27). In this area, the surface temperature reaches the necessary trigger temperature, near-surface convergence exists and the inhibition is low, CIN $\sim$ -10 J kg^{-1} and CAP $\sim$ 0 to 0.2 °C, therefore, clouds can form more easily than elsewhere (Fig.8.35). The KO-index shows a significant horizontal difference ranging from potential instability, KO $\sim$ -6 K, in the east to almost KO $\sim$ 0 K in the west, indicating a potentially stable atmosphere. Almost no CAPE $\leq$ 50 J kg^{-1} and LSI $\leq$ -2 °C, the latter indicating that this area is strongly capped by a mid-tropospheric inversion, Lid effect $\sim$ 2.5 to 3 °C (Fig.8.34), will prevent thermals from the boundary layer to penetrate into the free atmosphere. Thus, the low CAPE values and the highly capped mid-troposphere do not allow the evolution of deep convection in the area.

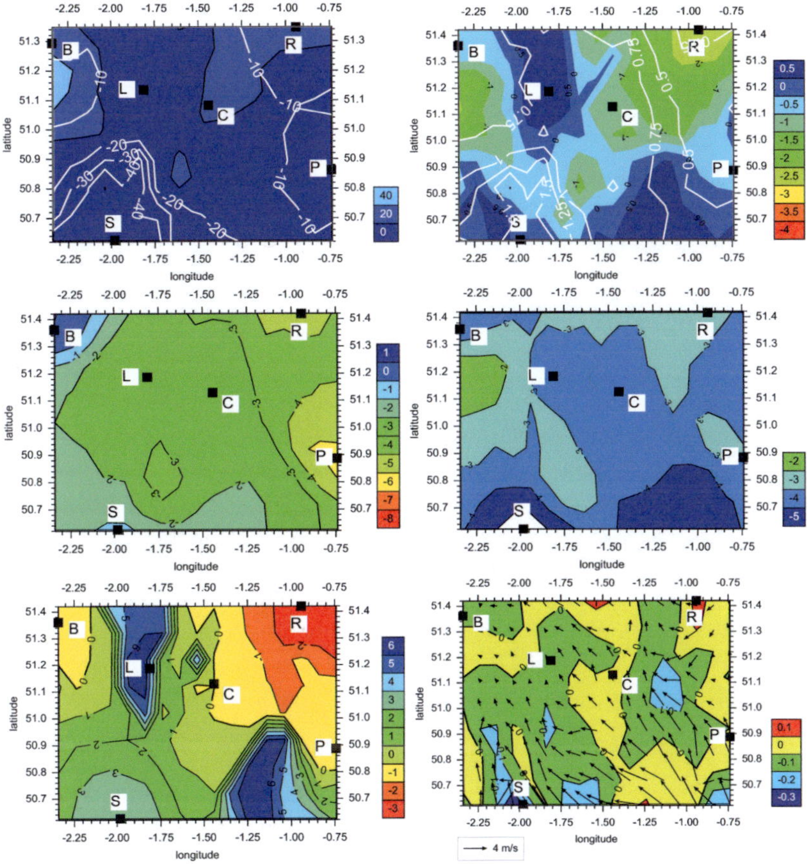

Figure 8.35: Combination of convection related indices at 1200 UTC on 18 August 2005. a) CAPE in J kg^{-1} (colour code) and CIN in J kg^{-1} (isolines) (top-left), b) LI in °C (colour code) and CAP in °C (isolines) (top-right), c) KO-index in K (center-left), d) LSI in °C (center-right), f) Tc-Tv in °C (bottom-left), and g) horizontal surface wind divergence x 10^3 s^{-1} (bottom-right).

At 1700 UTC (Fig.8.36) a remarkable increase in CAPE $\sim$ 700 J kg^{-1} is observed in the area between Bath and Larkhill radiosonde stations as a consequence of the significant increase in moisture in the PBL in the period from 1200 UTC to 1700 UTC (Fig.8.30). In this area, absolute values of CIN lower than 20 J kg^{-1} and CAP ≤ 1 °C as well as LSI ~ 2 °C indicate that atmospheric inhibition is low in the low- and mid-troposphere so that clouds can form not finding any mid-tropospheric lid constraining their vertical evolution. KO-index of approximately -8 K in association with the large-scale lifting of about -10 to -12 hPa h^{-1} (Fig.8.26) and the near-surface convergence are responsible for the erosion of lids and triggering in this area, resulting in deep convection.

One may conclude that isolated showers developed in this area as a consequence of the strong advection of moisture in the area between Bath and Larkhill, with a subsequent increase in instability, the weakening of the lid in this western region as a result of large-scale lifting, which at that time not longer exerts any constraining effect on the evolution of deeper clouds in the area, and near-surface and PBL convergence, which allow parcels to be lifted to their LFC since the convective triggering temperature was not reached at that time (Figs.8.33, 8.34). The conclusions drawn from the combined analysis of convection-related indices agree with the convection activity observed on that day. The early phase of the clouds that developed into these showers in the area between Bath and Larkhill is shown by a set of CAMRa RHIs along 255 and 244 degrees (highlighted in Fig.8.27). Mantle echoes with tops at 4 km due to convection penetrating the lid at 3 km are observed at about 1516 UTC (Fig.8.37). The cirrus canopy associated with the advance of the cold front from the west can be seen above 8 km. Forty minutes later, the tops of the convective clouds are still at 4 km (Fig.8.38). By 1630 UTC, an area of significant showers can be seen on the radar network at the eastern edge of the mesoscale area of brighter cumulus clouds. The RHI in Fig.8.38 shows that by 1633 UTC, the shower tops were up to almost 6 km. These showers continued to intensify while still remaining isolated, as shown by the radar display for 1700 UTC.

Summary and conclusions

IOP16 can be characterized as a warm and sunny day when locally initiated convection takes place. The synoptic-scale dynamics result in lifting in the area of about -2 to -8 hPa h^{-1}. Strong mid-tropospheric lids are present in the area, at 3 km and 6 km. Their temporal evolution and eventual erosion control the evolution of clouds further in the atmosphere and their transition from shallow to deep convective clouds.

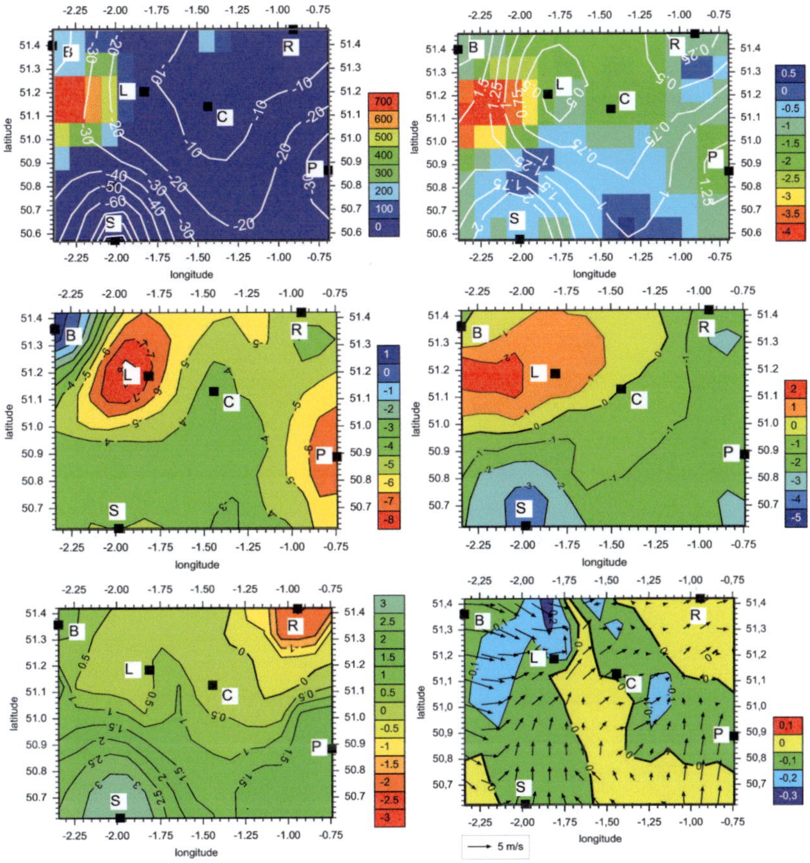

Figure 8.36: Combination of convection-related indices at 1700 UTC on 18 August 2005. a) CAPE in J kg^{-1}(colour code) and CIN in J kg^{-1} (isolines) (top-left), b) LI in °C (colour code) and CAP in °C (isolines) (top-right), c) KO-index in K (center-left), d) LSI in °C (center-right), f) Tc-Tv in °C (bottom-left), and g) horizontal surface wind divergence x 10^3 s^{-1} (bottom-right).

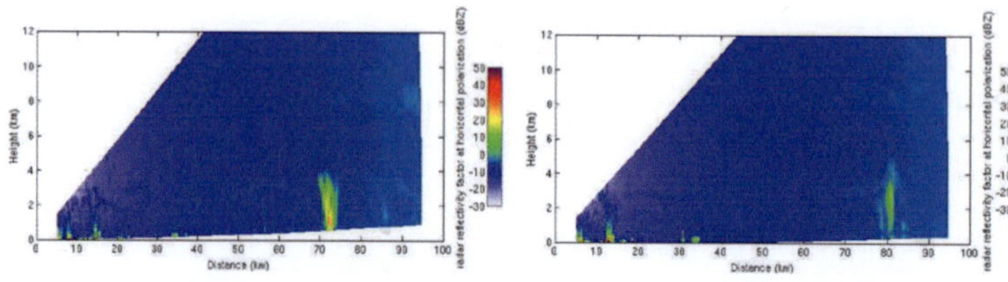

Figure 8.37: CAMRa RHIs at 1517 UTC (left) and 1519 UTC (right) along azimuths 255 and 244 degrees on 18 August 2005.

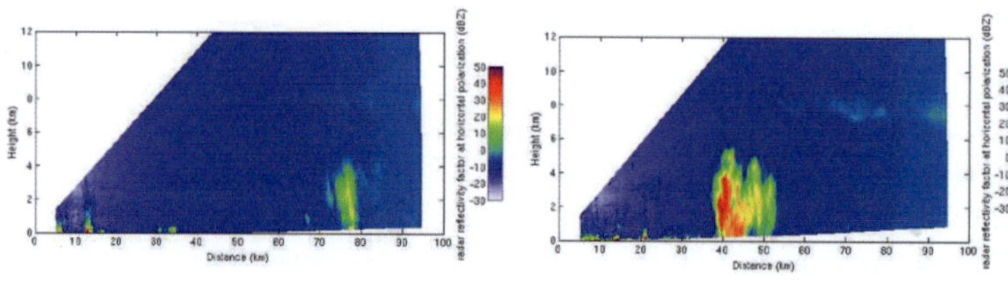

Figure 8.38: CAMRa RHIs at 1558 UTC (left) and 1633 UTC (right) along azimuths 244 and 283 degrees showing the radar reflectivity factor at horizontal polarization in dBZ (colour scale) on 18 August 2005.

It has been shown that two mechanisms cause the initial formation and posterior development of deep convective clouds:

- At about 1400 UTC, a low-level convergence line is established between Bath and Larkhill stations.
- In the period from 1500 to 1700 UTC, a strong increase in bounday layer moisture, due to the advection of moisture from the south-west, results in an increase in instability, CAPE, and a reduction in atmospheric inhibition, CIN.
- Where the low-level convergence zone coincides with the area of maximum large-scale lifting, the mid-tropospheric lids are eroded by adiabatic lifting of the profiles, and the warm moist air is lifted.

8.3.2 COPS IOP8b

Synoptic situation and evolution of deep convection

On the COPS IOP 8b, July 15, 2007, the COPS area is located in the transition zone between an eastern European ridge, stretching from the Mediterranean Sea to

Poland and a high amplitude eastern Atlantic trough (Fig.8.39). Near the surface, a shallow low develops, while at the 500 hPa level the ridge is still dominating. This leads to only little advection of positive vorticity in the upper troposphere and warm air in the lower troposphere throughout the whole day. As a consequence, the large-scale forcing results in weak lifting of 0 hPa h^{-1} to -6 hPa h^{-1} concentrating on the western part of the COPS area over the Vosges (Kottmeier et al., 2008).

Near the surface, the investigation area was in the transition zone between a surface low in the west and a surface high in the east. During the day, the surface low moved eastwards, accompanied by a decrease in surface pressure by about 4 hPa between 0600 and 1800 UTC. These synoptic conditions affected flow conditions in the PBL. In the morning, low-level south-easterly wind dominated in the COPS domain while in the afternoon south-westerly winds prevailed in the PBL.

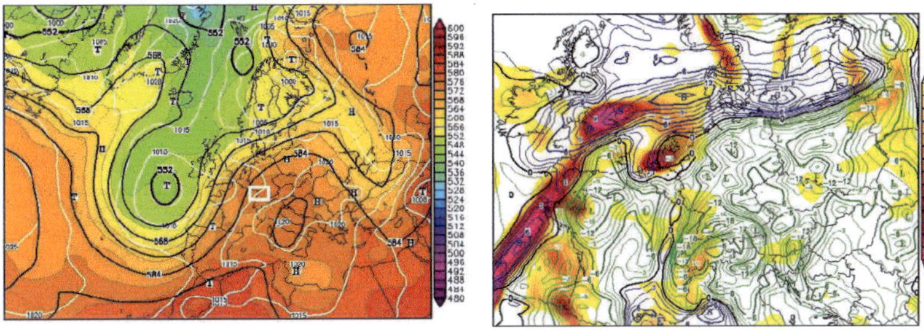

Figure 8.39: GFS analysis of surface pressure in hPa (white isolines), 500 hPa in gpdm (black isolines) and relative topography in gpdm (colour code) (left), and KO-index in K (isolines) and 500 hPa vertical motion in hPa h^{-1} (colour code) (right) at 1200 UTC on 15 July 2007. The COPS area is indicated by the white and black rectangle, respectively.

On this day, the first cumulus clouds form at about 1115 UTC over the Black Forest. At about 1200 UTC, two convetive clouds could be observed in the satellite images over the Black Forest. One is located about 7 km east of Hornisgrinde at 48.65 °N, 8.25 °E and the other at 48.10 °N, 8.35 °E. These cells still exist at 1230 UTC (Fig.8.40a). By this time, an additional weak north-south oriented cloud band forms east of Hornisgrinde between 48.45 °N, 8.30 °E. At 1400 UTC (Fig.8.40b), the isolated northern cell east of Hornisgrinde merge with the northern cloud band in which the convective activity had intensified. Simultaneously, the southern cell had been extended to a cloud band. With time both cumulus cloud lines unite

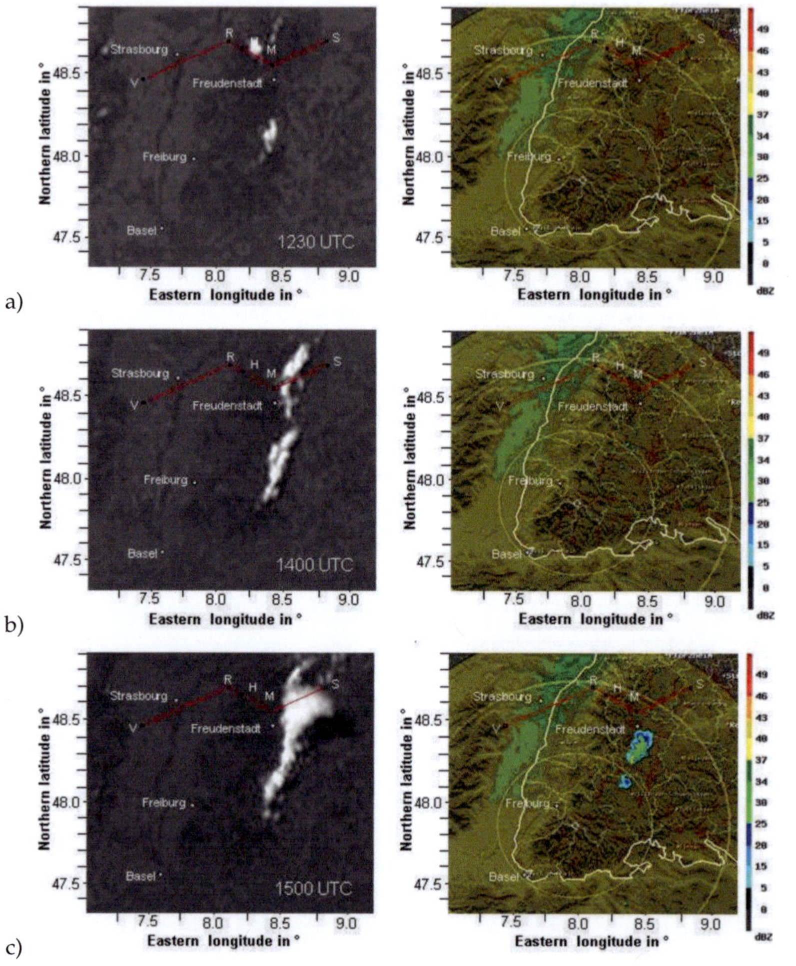

Figure 8.40: MSG satellite images (left) and low-level radar reflectivity (right) measured by the DWD radar positioned at feldberg in the southern Black Forest on 15 July 2007. Supersites are marked by black dots over the red line, from left to right; V(Vosges), R(Rhine valley/Achern), H(Hornisgrinde), M(Murgtal/Heselbach/AMF site), and S(Stuttgart/Deckenpfronn). The range rings increments in the reflectivity images are 47.5 km.

(Fig.8.40c) and at the center of the line, southeast of Freudenstadt, one cumulonimbus cloud (Cb) develops rapidly within 25 minutes up to 12 km height as seen from DLR radar (Fig.8.41) (Kottmeier et al., 2008). The reflectivity in the center of the Cb is high (60 dBZ), indicating local heavy precipitation with hail. The peak of the cloud development is at 1445 UTC.

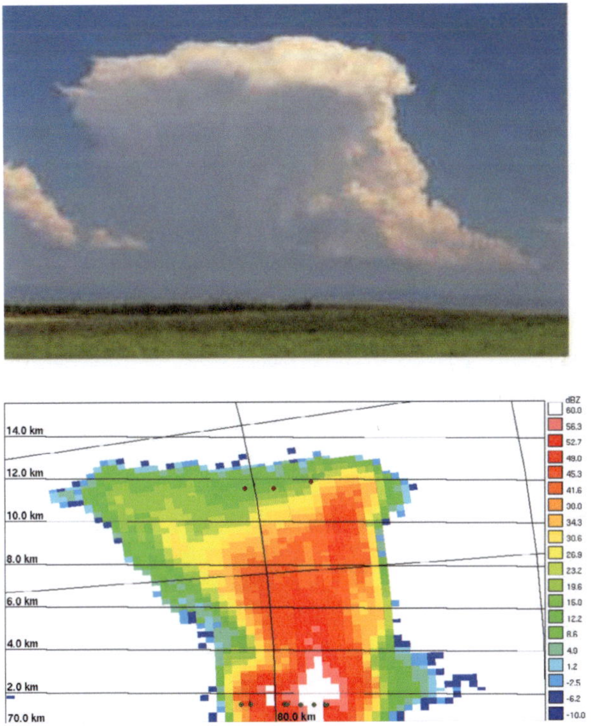

Figure 8.41: The cumulonimbus cloud over the northern Black Forest at the summit of its development, 1445 UTC, seen from the location of DLR-Poldirad radar in the Rhine valley (top). Reflectivity of the Cb-cloud given in dBZ (colour code) seen with DLR-Poldirad radar (bottom) (Courtesy of M. Hagen).

Near-surface and PBL conditions

To characterise the near-surface and the vertical structure of the atmosphere, near-surface observations and radiosoundings are used, respectively. Fig.8.42 shows the

near-surface temperature and dew point spatial distribution in the area of investigation at 0800, 1000, and 1300 UTC.
Early in the morning, by 0800 UTC, a temperature average of approximately 25 °C is observed in the area. Higher temperatures, around 28 °C, are found close to supersite H, Hornisgrinde. The area in the west is wetter with dew point temperatures of about 17 °C, than that in the east with dew point temperatures $\sim$ 12-14 °C.

Due to cloud-free conditions with high insolation, at 1000 UTC the mean temperature in the area has increased to about 28 °C. A strong increase in dew point, from $\sim$ 12-14 °C to $\sim$ 16-20 °C, is observed in a N-S oriented line in the area of supersite A, Achern, and supersite H coinciding with the area where the first cumulus clouds form at about 1115 UTC over the Black Forest. In the period from 0945 to 1100 UTC, wind direction in supersite H turns from south-east to west-south-west contributing to the transport of moist air from the west to the east. Surface measurements show an increase in surface-specific humidity from approximately 9 to 11.5 g kg^{-1} (Kalthoff et al., 2009).

The analysis of the energy balance components at the Hornisgrinde mountain peak (Fig.8.45) shows that from 1000 to 1100 UTC the latent heat flux reduces from $\sim$ 350 W m^{-2} to 300 W m^{-2} due to the moisture increase in the area. In addition and favoured by the high insolation, thermally induced wind systems developed over the Black Forest transporting moist air to the mountains and leading to low-level convergence over the mountain crests (Fig.8.42). By 1300 UTC, temperature has increased in the area by about 2 °C and most of the stations show temperatures over 30 °C. Dew point temperatures are similar to those at 1000 UTC except in the area of supersites A, H and M, Heselbalch (AMF site), where values of about 18 to 22 °C are measured. At supersite M, the near-suface wind direction changes in the period from 1300 to 1400 UTC from south-east to west (Fig.8.44d), advecting moist air into this area whereas in the west drying occurs (not shown). Fig.8.43 displays the radiosonde profiles at supersite A, in the Rhine valley, between 0800 and 1400 UTC, and supersite H, at the top of the mountain crest (northern Black Forest), for the same period of time. At Achern, besides warming in the PBL in the diurnal cycle, the whole troposphere is very dry, with shallow intermediate moister layers between 870 hPa and 700 hPa as well as between 530 hPa and 300 hPa. At 950 hPa, a PBL-capping inversion is observed, which at 1400 UTC has been effectively removed by the increase in boundary layer temperature. Humidity in the PBL increases from 0800 to 1100 UTC below 700 hPa, but from 1100 to 1400 UTC it does not change significantly, and in the layer from $\sim$ 875 hPa to 675 hPa it decreases. At supersite H, a small increase in temperature is observed for the whole period.

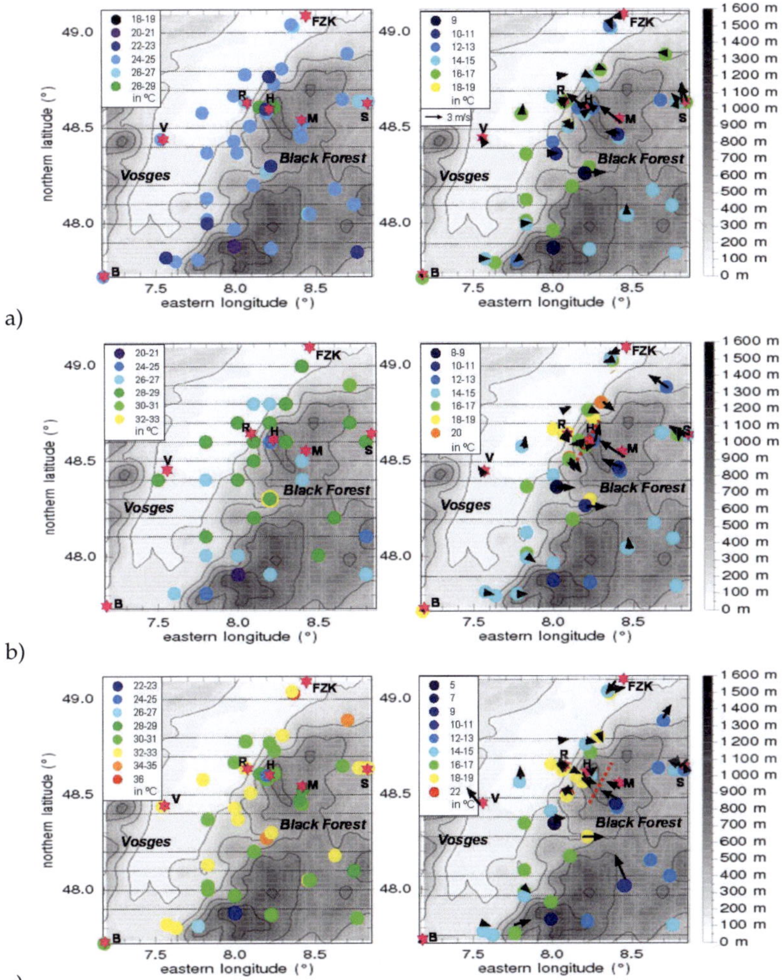

Figure 8.42: Topography of the COPS investigation area (black-white colour scale), near-surface temperature (colour code) (left), dew point temperature (colour code) (right) in °C, and wind vectors in m s^{-1} (arrows) at a) 0800 UTC, b) 1000 UTC and c) 1300 UTC on 15 July 2007. Supersites are marked by pink stars, from left to right; B (Burnhaupt le Bas), V(Vosges), R(Rhine valley/Achern), H(Hornisgrinde), M(Murgtal/Heselbach/AMF site), FZK (Forschungszentrum Karlsruhe) and S(Stuttgart/Deckenpfronn). The red dotted lines indicate the convergence zones along the mountain crests.

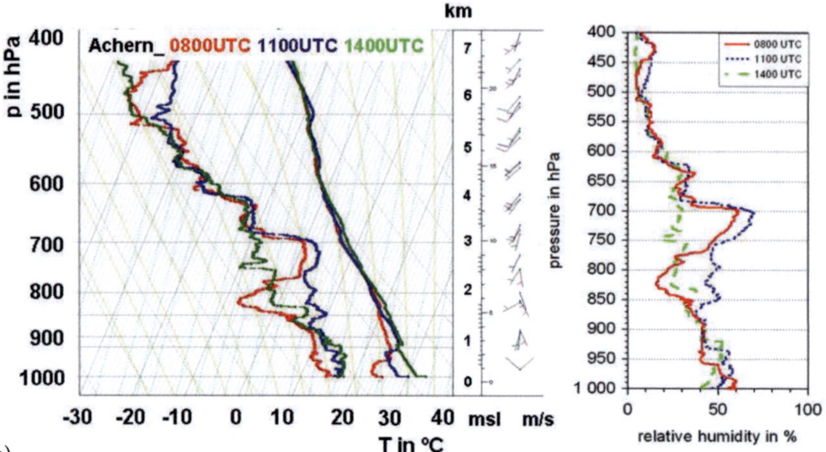

a)

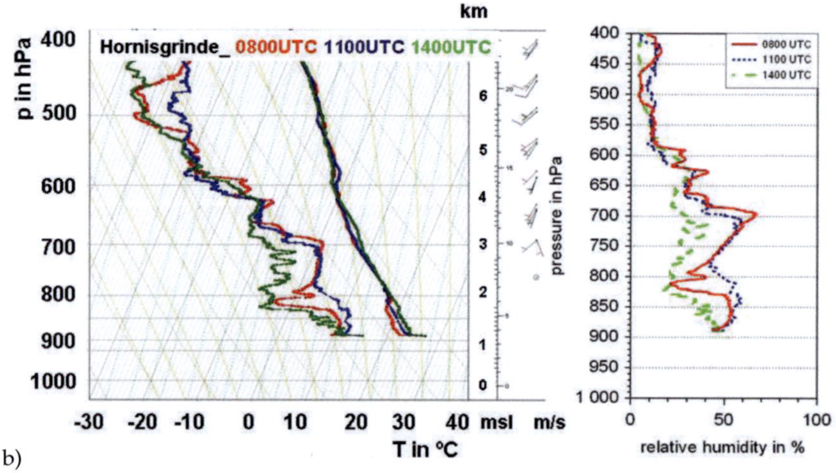

b)

Figure 8.43: Skew T-log p diagrams showing the temporal evolution of radiosonde profiles at a) Achern, and b) Hornisgrinde, and wind in m s^{-1} (half-barb=2.5 m $^{-1}$; whole-barb=5 m $^{-1}$)(left) and relative humidity in % (right), at 0800 UTC (red lines), 1200 UTC (blue lines) and 1400 UTC (green lines).

In the period from 0800 to 1100 UTC, moisture increases up to $\sim$ 780 hPa. At 1400 UTC, an important decrease in moisture takes place below $\sim$ 650 hPa. On that day, high insolation results in very high PBL temperatures. The warm but very dry air masses do not allow for early or widespread convection to develop. In addition, in all radiosonde stations in the investigation area a distinct PBL-capping inversion, a weaker inversion at about 2 km and some minor mid-tropospheric inversions existed. As already mentioned, in the morning low-level south-easterly wind dominated in the COPS domain while in the afternoon south-westerly winds prevailed in the PBL.

Making use of the available radiosonde profiles at supersites A and H, height-time cross-sections are interpolated. At supersite M, radiosonde profiles are only available at 1100 and 1700 UTC, however, cross-sections of absolute humidity in g m^{-3} and temperature in K are available from the scanning Humidity and Temperature Profiler HATPRO (Rose et al., 2005), (Fig.8.44), which allows investigating the conditions east of the Black Forest summit. At Achern, radiosonde profiles are available at 0800, 1100 and 1400 UTC. The station is about 150 m amsl. In the period from 0800 to 1100 UTC, mixing ratio increases from 10 to 12 g kg^{-1} below $\sim$ 500 m amsl. At 1100 UTC, mixing ratio of about 5-6 g kg^{-1} is observed up to $\sim$ 3300 m amsl, which is an indication of some convection activity in the area. At 1400 UTC, 5 g kg^{-1} are observed up to 1000 m amsl, and the depth of the 12 g kg^{-1} layer has not changed. At Hornisgrinde, radiosondes are available at 0800, 1100, 1400, and 1700 UTC. The station is about 1160 m amsl. From 0800 to 1400 UTC, the mixing ratio at the lower levels is approximately 9 g kg^{-1}. At 1100 UTC an increase of about 1 g kg^{-1} is observed in the layer from about 1200 m to 1500 m amsl. At Heselbach (500 m amsl), in the period from 1100 to 1330 UTC a strong increase in moisture up to 12 g m^{-3} in the lower levels takes place. The maximum increase occurs in the period from 1230 to 1330 UTC up to 3 km AGL, in contrast with the maximum increase observed at A and H supersites at about 1100 UTC. From this time, absolute humidity drops about 3 g m^{-3} in less than one hour. In the period from $\sim$ 1245 to 1400 UTC, the area is covered by clouds (Figs.8.45, 8.46) whose base is at about 3000 m AGL observed from the ceilometer located at this site (not shown). This reduces the total radiation from 600 to 450 W m^{-2}. From 1200 to 1300 UTC, latent heat flux decreases from $\sim$ 300 to 180W m^{-2}. This can be explained by the reduction in global radiation and the increase in moisture in the air (Kalthoff et al., 2009). In the period from 1300 to 1330 UTC, latent heat flux increases from about 180 W m^{-2} to 320 W m^{-2}.

Using a mean surface latent heat flux, E_0 of about 300 W m^{-2}, assuming $E_{z_i} \approx 0$ at the top of the PBL and a layer depth of nearly 1000 m, PBL-moistening in the order of 0.4 g kg^{-1} h^{-1} would be expected. Thus, advection of moisture plays an important role in the increase in moisture observed at both stations, Hornisgrinde and Heselbach, on that day. At Heselbach wind speed and wind direction can

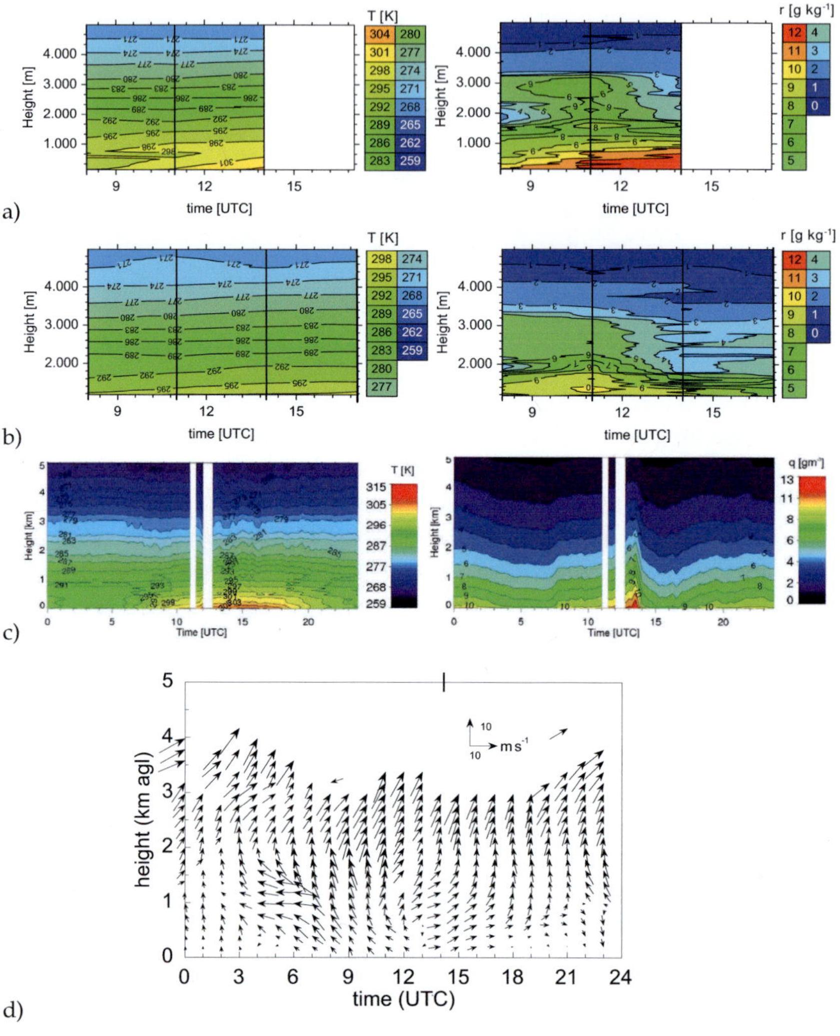

Figure 8.44: Height-time cross-sections of temperature (T) and mixing ratio (r) at a) Achern, and b) Hornisgrinde from interpolation of radiosonde profiles (indicated as black lines), and c) temperature and absolute humidity at AMF (Heselbach) from HATPRO system. d) Height-time cross-section of wind vectors measured using the wind profiler at the AMF site on 15 July 2007.

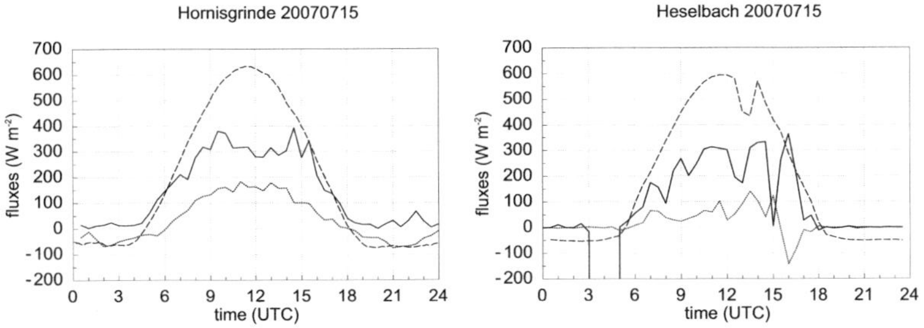

Figure 8.45: Energy balance components at Hornisgrinde (left) and Heselbach (right). Global radiation (dashed line), latent heat flux (solid line) and sensible heat flux (dotted line) in W m^{-2}.

be analyzed with a resolution of 1 hour from the radar wind profiler located at that station (Fig.8.44d). From 1300 to 13300 UTC, the wind turns from south-east to west transporting moisture into this station resulting in the strong increase in moisture observed in this station (Fig.8.44). This change in wind direction occurs slower at the lower levels and is due to the passage of the mesoscale convergence zone which was formed between the high pressure system east of the COPS domain, where south-easterly wind prevailed, and a surface low to the west with south-westerly wind. Therefore, the transport of moist air from the west-south-west and high evapotranspiration result in the strong increase in boundary layer water vapour observed at supersite H at about 1100 UTC, and supersite M at about 1300 UTC.

Azimuth scans from the HATPRO system (Rose et al., 2005) give the opportunity to investigate the spatial variability of the water vapour and cloud field at Heselbach (Martin et al., 2006). Within the framework of COPS, the HATPRO profiler was deployed on the roof of the Atmospheric Radiation Measurement Program (ARM) Mobile Facility (AMF) in the Black Forest Murg Valley (Kneifel et al., 2008). Regular azimuth scans of IWV with 5° resolution in azimuth at fixed 30° elevation were carried out approximately every 15 min. IWV data are shown as anomalies from the given offset value. A combined analysis at the AMF site of elevation scans and azimuth scans allows to investigate when and how much the moisture increases (Fig.8.44), and in which direction this change occurs (Fig.8.46). At that station, a strong increase in moisture was observed in the period from 1100 to 1330 UTC, with the peak of increase between 1230 and 1330 UTC (Fig.8.44c). We have seen that a change in wind direction at that station from 1200 to 1400 UTC (Fig.8.44d) is responsible for the advection of moist air from the west. These processes can be identified from the analysis of the IWV anomaly (Fig.8.46). At 1005 UTC, cloud-free conditions and a homogeneous distribution of water vapour are observed. At about 1251 UTC, the first clouds appear in the area in a line with orientation N-S. Water vapour is higher also in this direction. By 1321 UTC, the cloud band is moving towards the east and the higher water vapour is now in the north-east. In the period from 1300 to 1500 UTC, the highest water vapour is observed in the east due to the advection of moisture from the westerly winds. At 1700 UTC, the sky is free of clouds and again a homogeneous distribution of moisture is observed in the area.

Horizontal distribution of moisture by means of IWV from GPS

The spatio-temporal evolution of atmospheric water vapour can be analyzed using the GPS-derived IWV in the COPS domain. When analyzing this data one may take into account that the COPS area is a complex terrain where height variations from 100 to 1200 m amsl are found on a small horizontal scale. For example, the supersite Achern is at about 150 m amsl whereas the supersite Hornisgrinde, about 22 km to the south-east, is at 1169 m amsl. As already demostrated, about 20 to 40 % of the water vapour in the atmosphere is below the 900 hPa, therefore, the IWV measurements from a station at 100 m amsl are not comparable with that from a station at 500 m amsl, such as supersite M, or from one at 1200 m amsl, such as supersite H. Direct comparisons may be made between the temporal variation of IWV at the different stations or the water vapour content in different atmospheric layers.

Fig.8.47 shows the temporal evolution of IWV in Meistratzheim (157 m amsl), Achern (140 m amsl), Hornisgrinde (1169 m amsl), and Heselbach (511 m amsl). Achern and Hornisgrinde show the same temporal evolution despite the difference of about 5 kg m^{-2} due to the difference in elevation of the stations. At both stations, the maximum IWV is observed in the period from 1100 to 1200 UTC, coinciding with the development of the cumulus clouds over the Black Forest (Fig.8.40). Meistratzheim, approximately at the same elevation as Achern, shows a completely different evolution. Between 1100 and 1200 UTC, minimum IWV from 18 to 20 kg m^{-2} are measured at that station. At 1100 UTC, Meistratzheim is 8.5 kg m^{-2} drier than Achern. From 1100 UTC, IWV starts to increase, reaching its maximum of 25 kg m^{-2} at 1500 UTC. The Heselbach GPS station, supersite M, is at about 500 m amsl, however the IWV measurements at that station are for the period from 1200 to 1300 UTC only 1 kg m^{-2} lower than the values obtained for Achern (140 m amsl). At 1400 UTC AMF is 3.5 kg m^{-2} wetter than Achern. The evolution of IWV in this station agrees with the evolution of the near-surface and PBL humidity as well as with the convection activity observed on this day (Fig.8.40).

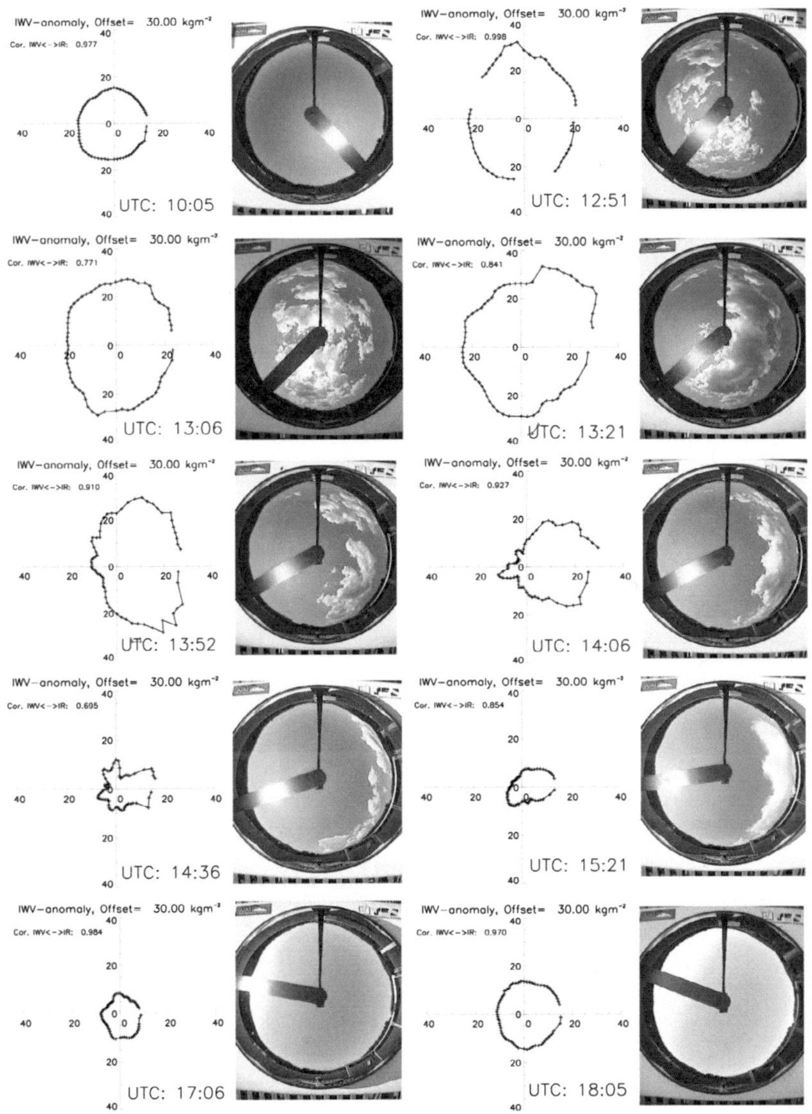

Figure 8.46: Azimuth scan at 30° elevation providing IWV in kg m^{-2} together with the corresponding sky picture of the Total Sky Imager (TSI) at the AMF site for the period from 1005 to 1805 UTC on 15 July 2007. Data are shown as anomalies from the given offset value. Courtesy of University of Cologne.

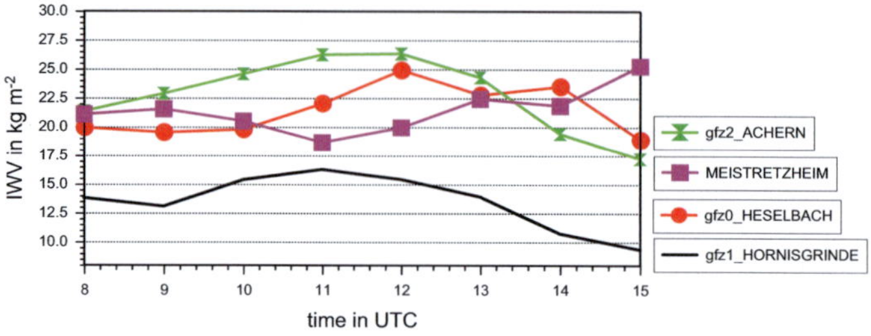

Figure 8.47: Temporal evolution of IWV in kg m^{-2} from GPS, at Meistratzheim, Achern, Hornisgrinde, and Heselbach from 0800 until 1500 UTC on 15 July 2007.

With respect to the CSIP GPS network, a much denser network is available in the COPS investigation area which results in a higher resolution of the IWV field. The GPS stations are mainly located in an east-west line around the supersites, however, in the north and south considerably fewer stations or even none in some areas are available. The spatial distribution of the IWV field is presented for the period from 0800 to 1500 UTC in Fig.8.48. For the interpolation, measurements from GPS stations outside the boundaries are used. By 1000 UTC, Fig.8.48, a line of high IWV from north to south is observed in the Black Forest. The maximum IWV at that time, $\sim$ 23-24 kg m^{-2}, is present north and south of Hornisgrinde and Achern coinciding with the areas where first cumulus clouds develop over the Black Forest at 1130 UTC. At 1200 and 1300 UTC, the highest IWV are measured in the same area. In the south, IWV is about 24 kg m^{-2} and in the north about 26 kg m^{-2}. By 1300 and 1400 UTC, IWV has decreased in the area of Hornisgrinde and Achern but has increased in the area between supersite M and supersite S because of the advection of moist air from the west-south-west (Fig.8.44d). This is in agreement with the surface and boundary layer analysis (Figs.8.42, 8.44c).

Deep convection likelihood

The use of IWV delivered from GPS has been proved to be very useful for detecting the location of primary initiated cells in an almost flat terrain as the CSIP investigation area. The main goal in this section is to test the usefulness of this method in a complex terrain such as the COPS domain.

In Figs.8.49, 8.50, the spatial distribution of the convection-related indices at 1100 UTC and 1400 UTC, respectively, is presented.

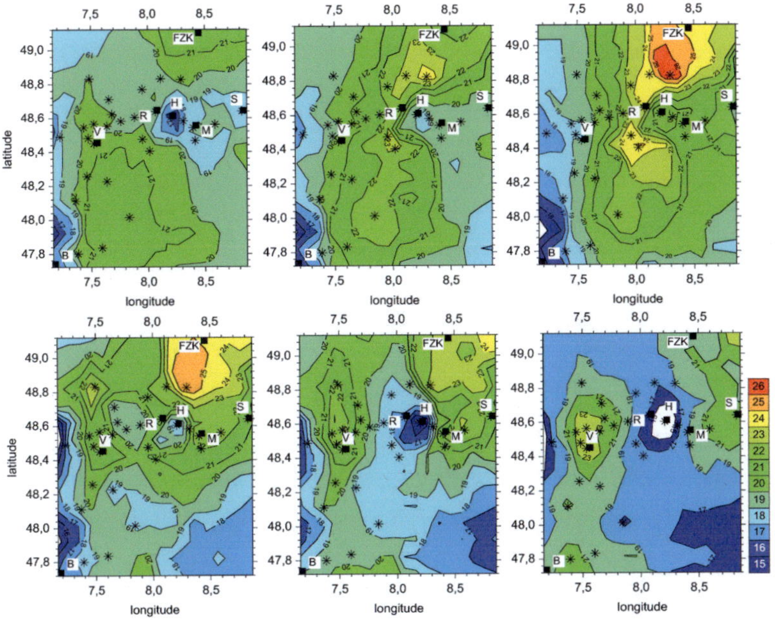

Figure 8.48: Spatio-temporal evolution of GPS-derived IWV in kg m^{-2} at 0800, 1000, 1100, 1300, 1400 and 1500 UTC (from top-left to bottom-right) on 15 July 2007. The positions of the radiosondes are indicated by black squares. GPS stations are marked by black stars.

At 1100 UTC, an area of low to moderate CAPE, 200-1000 J kg^{-1}, LI about -4 °C and KO about -8 K, is observed in a north-south oriented band positioned between supersites H and M. Over this band, two small areas of maximum CAPE $\sim$ 1000 J kg^{-1} and LSI ≥ 0 °C are observed, just in the area where the first cumulus clouds develop at 1130 UTC in the Black Forest. The latter indicating that convective clouds formed in these areas can penetrate further in the mid-troposphere without any mid-tropospheric inversions constraining its evolution. A significant horizontal difference in CIN, from -350 J kg^{-1} in the west of Hornisgrinde to -50 J kg^{-1} in the eastern region exists. The minimum -50 to -100 J kg^{-1} is observed in a north-south oriented line between supersites H and M, coinciding with the area of higher instability in that region. CAP indicates that at that time low boundary layer inversions are present in the whole area. In agreement with the CIN observations, maximum CAP is in the western areas, $\sim$ 5 °C, and minimum CAP in the vicinity of supersite H, $\sim$ 1.5 °C. Tc is not reached at this time, thus an additional mechanism is necessary to trigger convection. The existence of a boundary layer triggering mechanism to release convective potential energy in this area is discussed in the next section.

At 1400 UTC, CAPE $\sim$ 2000 J kg^{-1}, LI $\sim$ -7 °C and KO of $\sim$ -10.5 K, indicating large instability, have strongly increased, in the north-east, especially in the area between supersites M and S. In this area also LSI $\sim$ 0 °C is observed coinciding with the area where convective activity is maximum at this time. In this area, CIN in the order of -100 J kg^{-1} and CAP $\sim$ 2.5 °C make the evolution of new deep convective clouds difficult despite the fact that the convective triggering temperature is reached at supersite H.

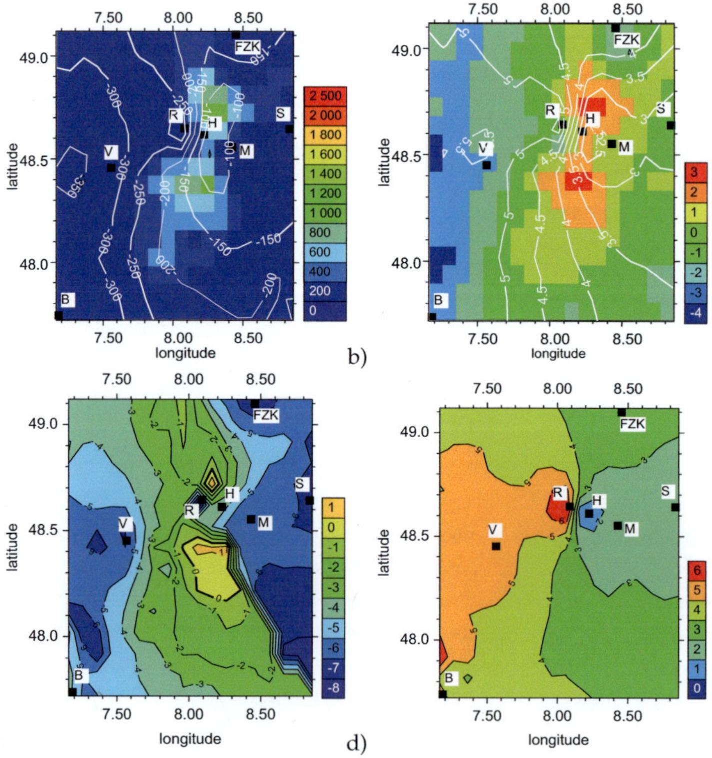

Figure 8.49: Combination of convection-related indices at 1100 UTC on 15 July 2007. a) CAPE in J kg^{-1}(colour code) and CIN in J kg^{-1} (isolines), b) LI in °C (colour code) and CAP in °C (isolines), c) LSI in °C, and d) Tc-Tv in °C.

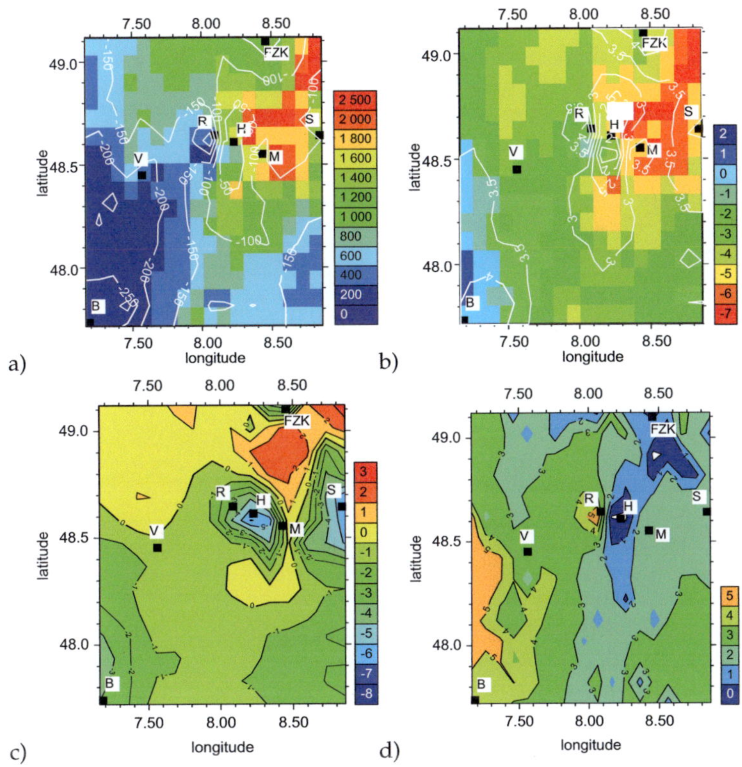

Figure 8.50: Combination of convection related indices at 1400 UTC on the 15 July 2007. a) CAPE in J kg^{-1}(colour code) and CIN in J kg^{-1} (isolines), b) LI in °C (colour code) and CAP in °C (isolines), c) LSI in °C, and d) Tc-Tv in °C.

Horizontal wind convergence

The analysis of the convection-related indices in Fig.8.49 correctly leads to the conclusion that initiation of convection in the COPS domain should take place in a N-S oriented line over the Black Forest. However, on this day, convection initiation in the COPS area is neither favoured by large-scale lifting (Fig.8.39), nor by surface-triggered buoyancy (Fig.8.49). Nevertheless, a line of convective clouds develops over the Black Forest. This line of convective clouds must be triggered by PBL-processes. As obvious from Fig.8.51 showing the line of sight Doppler velocity measured by the DWD radar on top of the Feldberg in the southern Black Forest, horizontal shear and convergence are observed in this area. As already mentioned, this mesoscale convergence zone was formed between a high pressure system east of the COPS domain, where south-easterly wind prevailed, and a surface low to the west with south-westerly wind. During the day the mesoscale convergence zone moved eastwards (Fig.8.39).

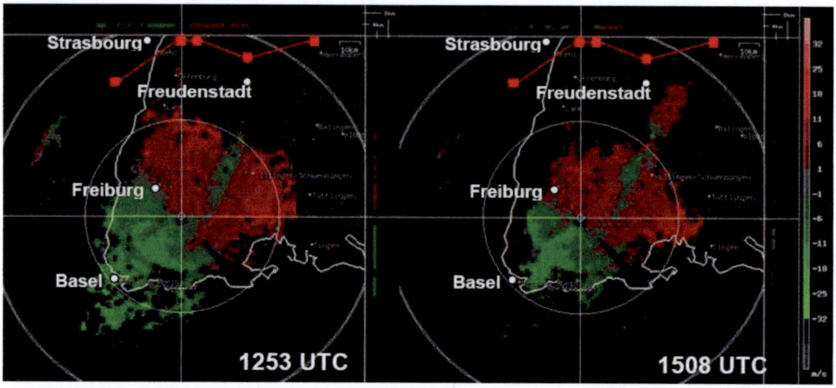

Figure 8.51: Doppler velocity in m s^{-1} at 1253 UTC and 1508 UTC measured and calculated from DWD radar on top of the Feldberg in the southern Black Forest on 15 July 2007. Green indicates a wind component to the radar and red gives wind away from the radar. The red squares indicate the positions of the supersites.

As obvious from the surface observations (Fig.8.42), a convergence line also exists along the mountain crests of the northern Black Forest at that time as a result of thermally induced wind systems (slope and valley winds) caused by high insolation. When the mesoscale convergence zone arrived to Hornisgrinde, the stationary thermally induced and the eastward moving mesoscale convergence superimposed leading to stronger vertical updraughts which penetrated the capping inversions reaching the LFC.

Summary and conclusions

The IOP 8b on July 15, 2007 is a locally initiated convection event. The convection is restricted to the Black Forest mountains area. The data analysis reveals that the convection is triggered by different mechanisms.

- Due to cloud-free conditions with high insolation, temperature and especially moisture increased within the PBL, which was reflected by an increase of atmospheric instability, indicated by an increase in the absolute values of

CAPE, LI, KO-index, and a reduction of inhibition, as indicated by CIN and CAP, as well as an increase of the LSI.

- The high insolation favours the development of slope and valley winds and high evapotranspiration results in the increase in moisture in the boundary layer and subsequent transport via the slope winds to the mountain crests. Aditionally, upvalley wind systems transported moist air to the mountains. Thus, advection of moisture plays an important role in the increase in boundary layer moisture in the area. The thermally driven circulation systems result in a convergence line along the mountain crest of the Black Forest (Kottmeier et al., 2008).
- Due to a high pressure system which is situated to the north-east of the Black Forest and a trough which approaches from the west towards the investigation area, a convergence line establishes in between.
- Both convergence lines optimally superimpose when the synoptic scale system reaches the Black Forest. The convergence in that area lifts the warm moist air upwards penetrating through the CBL-capping and upper level inversion. Consequently, CAPE is released and deep convection develops with heavy showers.

The spatial distribution and temporal evolution of atmospheric water vapour, highly related to the initiation and evolution of convection on this day, are adequately represented by the GPS-derived IWV. The combination of convection-related indices calculated from the combination of IWV from GPS and radiosonde data succsses-fully helps detecting the location of the initiation of deep convection in the area. Therefore, it can be concluded that the spatial resolution obtained by combining GPS and radiosonde data and the combination of convection-related indices investigated are also useful for the prediction of areas where deep convection likelihood is high in topographically complex terrains such as the COPS investigation area.

8.3.3 Conclusions about the comparison of the initiation of convection over flat and complex terrains

Comparison of the occurence of isolated thunderstorms in an almost flat terrain such as the CSIP domain and in a complex terrain such as the COPS domain, leads to the following conclusions:

- In both cases, warm and sunny conditions are present. Nearly cloud-free conditions and consequent high insolation leads to warm boundary layers. In the complex terrain this favours the evolution of secondary circulations.

- In the area where deep convection occurs, dry conditions in the morning do not allow early or widespread convection.
- Advection of moisture plays a very important role in convection initiation on both days. A strong increase in moisture is observed in a short period of time in the area where deep convection and subsequent showers develops.
- Moisture spatial distribution is more important than temperature distribution for the evolution of convection.
- The spatial distribution of IWV is highly related to the evolution of instability in the area.
- The combination of radiosonde data and GPS measurements, for the increase in resolution of the water vapour horizontal distribution, and the calculation of the spatial distribution of convection indices is a useful technique both over almost flat and complex terrains.
- In both cases, initiation of convection is neither favoured by large-scale lifting nor by surface-triggered buoyancy. Low-level convergence zones act as triggering mechanisms for lifting of the warm moist air upwards, eroding the existing lids and releasing CAPE.
- The combination of convection-related indices used in this thesis successfully detects the areas where deep convection is more likely.

Chapter 9

"Best Combination", Likelihood and Thresholds for the analysis of the Initiation of Deep Convection

Based on the CSIP cases investigated, it has been demonstrated: i) the need for detailed observations of the boundary layer thermodynamic and dynamic fields to better determine the timing and location of the initiation of convection, ii) the usefulness of the increase in resolution obtained using the synergy of different instruments to improve the representation and prediction of deep convection likelihood, and iii) the need of jointly evaluate several convection-related indices quantifying atmospheric stability, inhibition and triggering, to successfully detect the location of the initiation of deep convection. This combination of convection-related indices and the resolution obtained adequately represented the likelihood of the initiation of deep convection in all the events investigated.

Typical thresholds for these parameters were given in chapter 2. In general, these thresholds result from experiments performed in the United States, where convective events are mostly related to large convective instability and strong capping inversions, as demonstrated during the International H_2O Project (Weckwerth and et al., 2004). However, it is known that the performance of the convection indices is highly dependent upon the features of the areas of the world (Charba, 1979), which suggests the need for regionally specific studies. Due to the maritime British climate and the absence of major mountains, weak convective instability and capping inversions are observed on convective occasions. Therefore, the aim of this chapter is to find out whether the thresholds available in the literature are also useful in the southern United Kingdom for the prediction of deep convection likelihood or if they must be adjusted to the regional conditions. In addition, an attempt is made to find out a simple combination of convection-related indices for the prediction of deep convection.

9.1 Combination of convection-related indices and examination of thresholds

In this section, the convection-related indices obtained for all the CSIP-cases investigated are jointly evaluated to establish some general thresholds for the initiation of convection in the southern UK. 2-D plots of the convection-related indices calculated for all of the CSIP IOPs are presented in Fig.9.1. In three of the 2-D plots shown, an index representing atmospheric stability, either CAPE, LI, or B, is plotted against an index representing atmospheric inhibition, either CIN, CAP, or Lid effect. The fourth figure represents a possible triggering mechanism, which was evaluated by using the Tc-Tv difference and the near-surface horizontal divergence. The field of interpolated radiosondes after combining GPS and radiosonde measurements is used to obtain CAPE, LI, B, CIN, CAP and Lid effect, and the interpolated radiosonde field after combining GPS, radiosonde, AWSs, and SYNOP stations is used to obtain Tc-Tv and the near-surface divergence. According to the values representing events where deep convection is observed, areas where deep convection is likely are marked within striped rectangles. The new thresholds for the convection-related indices are the boundary values of these rectangles. Within these areas also events with no convection are observed. This confirms that it does not suffice to analyze single indices to correctly predict the likelihood of deep convection. Even the use of two indices, e.g. one representing atmospheric instability and the other showing atmospheric inhibition, does not accurately represent the likelihood of deep convection. According to the initial working hypothesis, convective storms are likely when the values obtained from the convection-related indices are contained within the four boxes in Fig.9.1.

The combination of CAPE and CIN (Fig.9.1a) indicates that convective episodes are limited to the area where CAPE is over 500 J kg^{-1} and absolute values of CIN below 50 J kg^{-1}. The combination of LI and CAP (Fig.9.1b) shows that deep convective events are limited to LI $<$ -1°C and CAP below 2 °C. Both LI and CAP are related with measures of CAPE and CIN, respectively. In the next section, the relation between these parameters is investigated. The combination of B and Lid effect (Fig.9.1c) shows the following threshold values B $<$ 0 °C and Lid effect $<$ 0 °C. The triggering of deep convection (Fig.9.1d) is controlled by convergence lines with strength higher than 0.1 x 10^{-3} s^{-1} and the condition Tc-Tv $<$ 0 °C is not a useful indicator for deep convection initiation since deep convective events are also observed for Tc-Tv > 0 °C. However, it is observed that higher values of the Tc-Tv difference imply the need of stronger low-level convergence for the initiation of deep convection to occur. Moreover, for Tc-Tv > 5 °C no deep convection is observed even though low-level convergence exists.

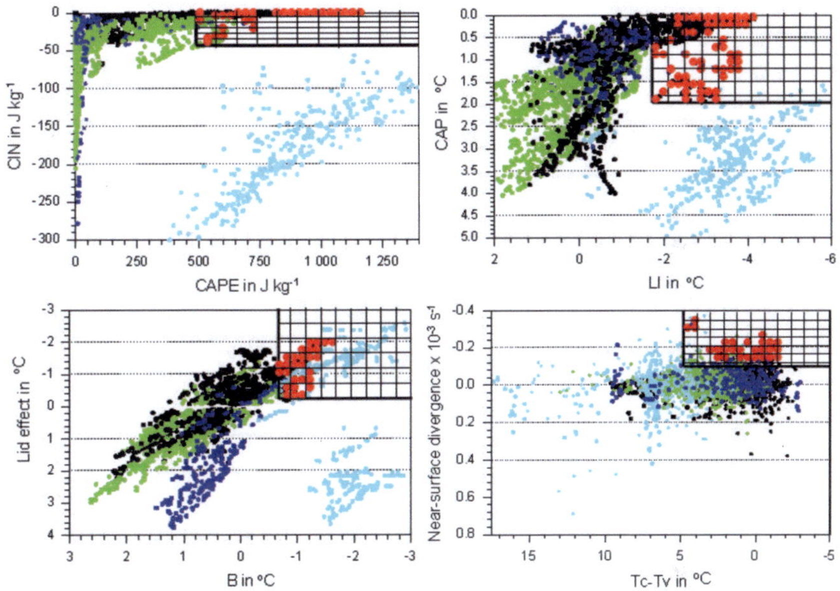

Figure 9.1: Combination of convection indices and examination of thresholds for the IOP2 (light blue), IOP5 (black), IOP8 (green), and IOP18 (dark blue). Each point represents a calculated value of the convection-related indices for a sounding. Red points represent the convection-related indices in areas where initiation of deep convection was observed (in this case no separation has been made concerning the different IOPs). Hatched rectangles indicate high deep convection likelihood. (Top-left) CAPE versus CIN in J kg^{-1}, (top-right) LI versus CAP in °C, (bottom-left) B versus Lid effect in °C, and (bottom-right) Tc in °C versus near-surface divergence x 10^{-3} s^{-1}.

The IOP2 (light blue) clearly shows the need of evaluating all of the aforementioned conditions to predict deep convection likelihood. In this case, deep convection does not occur in the area of investigation. Despite the moderate to high CAPE values, from 500 to 1400 J kg^{-1}, all radiosondes show CIN < -50 J kg^{-1} and strong

CAP. As previously discussed in section 8.1, the near-surface and boundary layer convergence observed in the area was not enough to overcome the strong CIN. In Tab.9.1, the thresholds obtained from this analysis are summarized,

CONVECTION-RELATED INDICES	THRESHOLDS
CAPE and CIN in J kg^{-1}	> 500 J kg^{-1} and > -50 J kg^{-1}
LI and CAP in °C	< -1 °C and < 2 °C
B and Lid effect in °C	both < 0 °C
Near-surface wind divergence	< -0.1 x 10^{-3} s^{-1}

Table 9.1: Thresholds for the convection-related indices calculated for the southern region of the UK.

When all these conditions are fulfilled the likelihood of deep convection is high. In comparison with the thresholds available in the literature, it can be stated that according to these thresholds the British climate leads to weak convective instability and capping inversions on convective occasions. One should note that the values presented in Tab.9.1 are from a limited data set and should be regarded as a guideline.

9.2 Instability and inhibition indices

To investigate the relation between CAPE and LI, as well as CIN and CAP, the observations from the CSIP IOPs are used. Fig.9.2 shows 2-D diagrams of LI versus CAPE values and CAP versus CIN values. In general, CAPE values over 500 J kg^{-1} correspond to LI values lower than -1 °C (Fig.9.2a) and absolute values of CIN smaller than 50 J kg^{-1} correspond to CAP values below 2 °C (Fig.9.2b) in agreement with the thresholds established in the last section where each of these indices was evaluated independently of its couple (Tab.9.1). However, e.g for a value of LI of approximately -1 °C, CAPE values in the range of approximately 100 to 600 J kg^{-1} are observed (Fig.9.2a). Also, for CAP of approximately 2 °C, CIN from -50 to -100 J kg^{-1} are found (Fig.9.2b). This suggests that threshold values of CAPE and CIN implicitely give the threshold value of LI and CAP, respectively, but not the other way round. This does not mean that the use of CAPE and CIN for evaluating the likelihood of deep convection is more appropiate. One should remember that it is the combination of CAPE-CIN and LI-CAP which is relevant for the evaluation of deep convection likelihood and not the individual performance of each index. As evident from Fig.9.1a,b and Tab.9.1, both combinations equally represent the probability of deep convection.

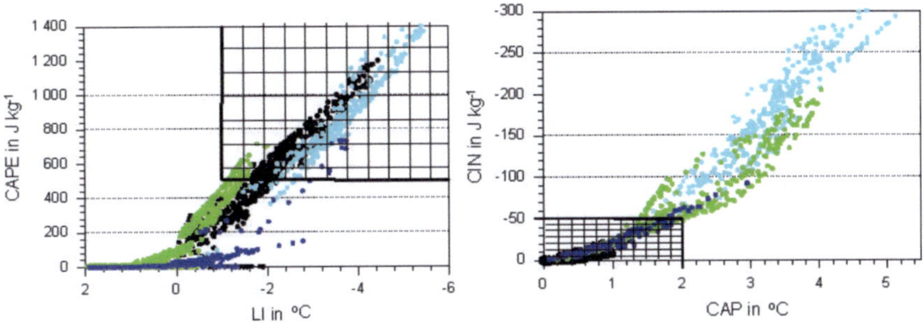

Figure 9.2: Relation between (left) CAPE vs LI and (right) CIN vs CAP. Observations from the IOP2 (light blue), IOP5 (black), IOP8 (green), and IOP18 (dark blue) are used. The boundaries of the boxes indicate the thresholds established for the prediction of deep convection for each index.

9.3 Best combination

In this section, an optimum and simple combination of convective indices is searched for the prediction of deep convection. The most promising index for this purpose is the LSI, Eq.2.31, since it combines the inhibitor effect of mid-tropospheric lids and atmospheric stability. However, this index represents neither the atmospheric inhibition at low-levels, up to here evaluated using the CAP index or CIN, nor the existence of a triggering mechanism.

To be able to consider atmospheric inhibition at low-levels, an extended concept of the LSI is introduced. The original LSI introduced by Carlson et al. (1980) consist of two terms, a) lid effect, and b) buoyancy. The lid term takes into account the inhibitor effect of mid-tropospheric inversions from 850 to 500 hPa, and the buoyancy term is a measure of the atmospheric conditional instability. The LSI is modified in this section in the lid term. The inhibitor effect of atmospheric temperature inversions is extended up to 50 hPa above the surface, thus, the inhibitor effect of capping inversions from 950 hPa to 500 hPa is taken into account. In this way, the information given by the analysis of CAP, CIN, and lid effect is compressed in one index.

It is necessary to include in the analysis some information about the existence of a triggering mechanism. As discussed in section 1 of this chapter, the analysis of Tc-Tv < 0 °C is not conclusive for the prediction of initiation of deep convection since many convective events occur for Tc-Tv > 0 °C, i.e., the convective triggering temperature has not been reached. However, in all cases horizontal wind convergence in the boundary layer played a significant role in the initiation of deep convection. The ability of low-level convergence to lift a parcel to its LFC highly depends on the strength of the convergence itself and on the amount of CIN. To take into account both factors, the ratio between the vertical wind velocity obtained from the horizontal wind divergence field, w_{CON}, and the vertical wind velocity obtained from the CIN field w_{CIN}, is calculated. The w_{CON} is an indication of the strength of the convergence and the w_{CIN} gives the updraught velocity needed to overcome the CIN. For example, for an absolute CIN = 50 J kg^{-1}, the equivalent vertical velocity of the updraughts to overcome the CIN is about $w_{CIN} \geq (2\text{CIN})^{0.5}$ = 10 m s^{-1} (Eq.2.29). In order to determine the resulting vertical wind in a convergence zone, the continuity equation is applied (Eq.2.17). It is assumed that the top of the PBL is at about 1 km. The analysis is restricted to the area where radiosondes are available. The radiosonde-interpolated wind field in the CSIP domain is used and the resolution at the near-surface is increased using the high-resolution wind measurements from AWSs and SYNOP stations. In this case, deep convective events are only observed within the CSIP domain on the IOP5 (black) and IOP18 (dark blue). Based on the above considerations, a 2-D diagram is introduced (Fig.9.3) where the LSI is plotted versus the ratio of the vertical wind velocity obtained from the horizontal wind divergence field, w_{CON}, and the vertical wind velocity obtained from the CIN field,

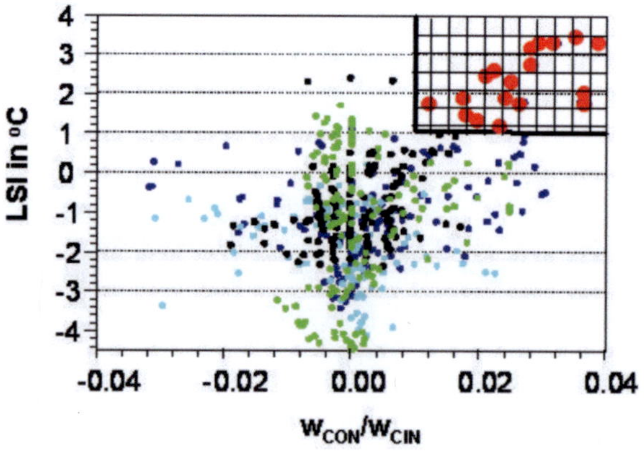

Figure 9.3: 2-D diagram for the prediction of deep convection likelihood. Extended LSI is represented versus the ratio w_{CON}/w_{CIN}. Observations from IOP2 (light blue), IOP5 (black), IOP8 (green) and IOP18 (dark blue) are used. The boundary of the box indicates the thresholds established for the prediction of deep convection.

w_{CIN}. Data points corresponding to areas where deep convection occurs are highlighted in red. A cut-off value of the extended LSI of about 1 °C and about 0.01 for the ratio w_{CON}/w_{CIN}, is found for the data set investigated. These values should be taken as a guideline since the uncertainties associated with these calculations are significant. Nevertheless, the positive value of the ratio, w_{CON}/w_{CIN}, suggests that all of the initiation of deep convective events were associated with boundary-layer convergence.

Chapter 10

Summary and Conclusions

The problem of forecasting convective precipitation is important because of the amounts of damage being caused by flooding as a result of the storms. A number of storms have recently caused substantial flooding and other damage in Europe. Such events are highly localized. Predicting where such storms will break out is one of the major challenges facing meteorologists. To the present day, operational numerical models often fail to predict the location and timing of convection initiation under weak or nonexistent synoptic forcing when triggering of convection results from boundary layer processes. These boundary layer processes are not well represented in the models, although boundary layer forcing has long been known to be a cause of thunderstorm initiation (Purdom, 1982). Therefore, a detailed knowledge of the thermodynamics and dynamics of the boundary layer is required to be able to predict the location of deep convective events. When only routine observations are available, the difficulties are referred to as being due to the lack of data.

In the last years, two important experiments were performed in Europe with the aim of gaining a better understanding of why deep convection breaks out precisely where and when it does; the Convective Storm Initiation Project (CSIP), performed in summer 2005 in the southern United Kingdom (Browning et al., 2007) and the Convective and Orographically-Induced Precipitation Study (COPS), performed in summer 2007 in the southwestern Germany and eastern France (Kottmeier et al., 2008). In this thesis, experimental data from both campaigns are used. The advantage of the CSIP and COPS campaigns was the dense observation networks available in the investigation area.

One goal of this thesis was to investigate whether the data obtained allow for the detection of the initiation and development of deep convection based on the analysis of the spatio-temporal evolution of convection-related indices which indicate atmospheric stability, available energy, convective inhibition, and triggering. The atmospheric stability is quantified using the Convective Available Potential Energy (CAPE), the Lifted Index (LI), and the KO-index. The atmospheric inhibition is investigated using the Convective Inhibition (CIN), the Capping inversion index (CAP) and Lid effect index. In addition, the Lid Strength Index (LSI), which combines a measure of the buoyancy (B) and the Lid effect is investigated. The existence of a triggering mechanism is evaluated calculating: i) the difference between the convective triggering temperature (Tc) and the near-surface virtual temperature (Tv), and ii) the near-surface horizontal wind divergence.

High-resolution aircraft measurements from the Do-128 flight along a 36 km track (approximately 10 min flight time) at about 510 m amsl were used to investigate the temperature and moisture variability in the boundary layer between the radiosonde stations. The analysis for the IOP5, 29 June 2005, showed a temperature variation of about 1.5 °C and of mixing ratio of about 2.5 g kg^{-1} along this flight track. Using a radiosonde profile in the area and taking first the maximum and second the minimum temperature and moisture measured along the flight track as representative of the boundary layer conditions, a difference in CAPE of approximately 680 J kg^{-1}, a difference in CIN of ~-150 J kg^{-1}, ~5.1 °C in LI, and ~2 °C in CAP is found. Thus, the atmospheric conditions vary from low to high probability for deep convection. When using the standard deviations, the atmospheric conditions vary from low probability (CAPE = 70 J kg^{-1} and CIN = -36 J kg^{-1}) to moderate probability (CAPE = 302 J kg^{-1} and CIN = -1 J kg^{-1}) for deep convection. The boundary layer variability, especially significant in the moisture field, strongly influenced the spatial distribution of convection indices, thus, the representation of the atmospheric conditions obtained from them. Although the spatial resolution given by the radiosondes was high, 25 to 50 km for CSIP, it was not enough to adequately represent the boundary layer variability on the scales relevant for the initiation of convection. Therefore, for a successful detection of convective initiation, an enhancement of the spatial resolution given by the radiosondes network was necessary. Since in general convection indices are calculated lifting a parcel from the near-surface or from a representative level of the boundary layer, an increase in resolution at the near-surface and in the boundary layer turned out to be particularly useful.

To achieve an optimal resolution, the synergy of different instruments was used. This was done combining the radiosonde data with the high resolution measurements of the network of automatic weather stations (AWSs) and synoptic (SYNOP) stations to increase the resolution at the near-surface, as well as global positioning system (GPS) measurements to increase the resolution of the water vapour field, thus, the boundary layer moisture distribution. Since the GPS measurements are retrievals of column-integrated water vapour mixing ratio, an algorithm, "adjustment method", was introduced to relate the GPS-derived integrated water vapour (IWV) measurements to the radiosonde profiles. Using the higher-resolution temperature, moisture and wind fields, the spatio-temporal distribution of convection-related indices was determined. Convection indices were computed lifting a parcel with near-surface characteristics, surface-based (SB) parcel, or averaged over the lowest 50 hPa of the PBL, 50 hPa mixed layer (ML) parcel to analyze whether the parcels are - with respect to the observation of deep convection - better defined in terms of the near-surface or averaged over the lowest 50 hPa of the PBL. The analysis of the convection-related indices and the effect that an increase in resolution had on their spatial distribution demonstrated that it is not enough to analyze one single parameter to correctly predict deep convection likelihood. Therefore, a working hypothesis was established for the prediction of areas where deep convection likelihood is high: "the prediction of deep convection should be limited to regions of high conditional instability, low atmospheric inhibition, absence of mid-tropospheric lids and the existence of a triggering mechanism". Thus, an important goal was to verify this hypothesis and, if possible, to define a combination of convection-related indices to better represent deep convection likelihood.

Five Intensive Operational Periods (IOPs), four from the CSIP campaign and one from the COPS campaign were investigated to evaluate the usefulness of an increase in resolution and the performance of the above-mentioned working hypothesis under different atmospheric conditions. Events with non-existent or weak large-scale lifting were chosen to investigate the initiation of deep convection via boundary layer processes. Observations from the initiation and development of an isolated storm were compared using a CSIP and a COPS case to establish the similarities and differences in the processes leading to deep convection over flat (CSIP) and complex (COPS) topographic terrains. Since regional conditions have been proved to exert a significant influence on the performance of convection indices on convection episodes (Charba, 1979), all observations from the CSIP IOPs investigated were jointly evaluated to ascertain whether the thresholds described in the literature for the analysis of convection indices and deep convection likelihood are applicable to the southern area of the United Kingdom (UK). Finally, to simplify the prediction of deep convection, an attempt was made at finding out the "best combination" of convection-related indices for the southern area of the UK.

The major conclusions and findings of this investigation can be summarized as follows:

- Although the spatial resolution of radiosondes turned out not to be enough to adequately represent the significant variability of temperature and water vapour in the boundary layer, radiosondes remain a useful and necessary tool for evaluating the spatio-temporal evolution of capping inversions or lids, and the vertical distribution of water vapour in the atmosphere. Using radiosonde profiles, it was demonstrated that, i) the estimation of the local potential temperature and humidity change in the PBL at the radiosonde sites is useful for evaluating the evolution of capping inversions, which have been proved to be relevant for the prediction of the location of initiation of convection, and the evolution of convective indices, and ii) 30 to 40% of the atmospheric water vapour is below the 900 hPa level. Therefore, the total-column atmospheric water vapour variability is dominated by the boundary layer variations.

- The use of the high-resolution near-surface observations allowed a detailed localization of convergence zones. These convergence zones were highly related to areas where deep convection was initiated even in cases where low atmospheric inhibition was observed. However, the observations also proved that convergence zones must not inevitably trigger moist convection, such as for the CSIP-IOP2 where high CAPE and PBL-convergence were present but high CIN suppressed convection, or on the CSIP-IOP8 where the convergence lines generated by the sea-breezes, developed from all coasts, were not enough to trigger deep convection.

- The increase in resolution obtained by the combination of radiosondes and GPS measurements resulted in a better spatial representation of the atmospheric conditions leading to deep convection. Even in regions without radiosonde data the use of GPS measurements was especially helpful in establishing the conditions in the area.

- Events with low CIN in the whole area at the time where the initiation of convection took place showed that smaller contour intervals than the typical 50 J kg^{-1} are required to accurately forecast the location of primary initiated cells. The additional analysis of CAP was found to be very useful.

- The determination of the altitude, strength, and evolution of low-level and mid-tropospheric lids was of key importance in order to accurately analyse the location and timing where inhibition existed or diminished, respectively.

- From this investigation it can be concluded that storm initiation depended more on boundary-layer convergence and low CIN than on high CAPE values. Location and timing of the initiation of convection were critically influenced by the structure of the humidity field in the planetary boundary layer. Additionally, the initiation of deep convection was better represented by indices calculated with parcels characterized by mixed-layer conditions, rather than with a parcel lifted from the near-surface. Once triggered, the presence of moderate to high CAPE and a high amount of mid-tropospheric moisture contributed to the growth of the initiated convective cells into deep-convective ones.

- The analysis of the spatio-temporal evolution of convection-related indices showed that the likelihood of deep convection cannot be predicted correctly by analyzing only one parameter. The resolution obtained and the combination of indices presented successfully detected the areas where deep convection likelihood was higher, both over flat (CSIP) and complex (COPS) terrains and under different atmospheric conditions. High-resolution water vapour profiles, boundary-layer convergence, and an accurate representation of CAPE, CIN and mid-tropospheric lids (quantified by the lid effect index) were needed to specify the location of convective initiation.

In addition, from the joint evaluation of all CSIP IOPs investigated, some thresholds were established for the combination of convection-related indices for the CSIP region. Deep convection was observed in areas where CAPE is higher than 500 J kg^{-1}, absolute values of CIN are below 50 J kg^{-1}, LI$<$-1°C, CAP below 2°C, lid effect$<$0°C, and near-surface convergence with strengths lower than -0.1 x 10^{-3} s^{-1}. These thresholds confirm that convective instability and capping inversions in that region are often quite weak on convective occasions and triggering is controlled by boundary layer convergence lines. These thresholds should be regarded as a guideline because of the uncertainties in the calculations and the limited number of IOPs investigated.

To simplify the prediction of deep convection a 2-D diagram is introduced where the LSI is plotted vs the ratio between the vertical wind velocity obtained from the horizontal wind divergence field and the vertical wind velocity obtained from the CIN field, w_{CON}/w_{CIN}. This combination takes into account all the conditions required for the occurrence of deep convection and allows the detection of areas with high likelihood of deep convection. A cut-off value of the extended LSI of about 1 °C and about 0.01 for the ratio w_{CON}/w_{CIN} is found. These values should be taken as a guideline due to the uncertainties in the calculation of these parameters.

Convection initiation is a vast and complex problem, but the synergetic use of different instruments to obtain a high-resolution representation of the boundary layer thermodynamic and kinematic fields has been proved to enhance the detection of the location of the initiation of deep convection. This study highlights that numerical weather prediction models should be able to predict the location of convergence zones and the altitude, strength and persistence of lids so that accurate forecasts can be made of the location and timing of the initiation of convective precipitation. Furthermore, it has been demostrated that accurate, high-resolution, three-dimensional moisture measurements are critical for the forecast of precipitating convection initiation. It has been shown that the GPS technology has great potential for this application. Moreover, further research is under way to determine how the assimilation of these data into weather prediction models affects the precipitation forecast. Some first results suggest a clear positive impact on the short-range precipitation forecast as well as the forecast of the water vapour field (Yan et al., 2009). For future experiments, it is highly recommended to use a homogeneous distribution of GPS stations rather than an accumulation of several GPS stations in specific areas. GPS stations every 10 km would duly represent the atmospheric water vapour distribution. Moreover, it is recommended to install at least some GPS and radiosonde systems at the same sites so that a systematic bias between both instruments can be determined.

List of Figures

List of Tables

List of Symbols

B	Buoyancy acceleration
c_p	Specific heat of dry air at constant pressure
c_p	Specific heat of water vapor at constant pressure
c_v	Specific heat of dry air at constant volume
c_{pd}	Heat capacity at constant pressure for dry air
c_{vd}	Heat capacity at constant volume for dry air
c_{pv}	Heat capacity at constant pressure of water vapor
c'_p	Effective heat capacity at constant volume for mixture of dry air and water vapor
c'_v	Effective heat capacity at constant pressure for mixture of dry air and water vapor
c_l	Heat capacity of liquid water
d	Grid distance
e	Vapor pressure
f	$=2\Omega\sin\Phi$ Coriolis parameter
g	Gravity or gravitational acceleration
H	Scale height
H	Relative Humidity
L_v	Latent heat of vaporization
N	Buoyancy or Brunt-Vaisala frequency
p	Pressure
p_0	Standard pressure
p'	Perturbation pressure
p_{00}	Standard pressure ($\approx$1000 hPa)
r	Mixing ratio
r_t	Total water mixing ratio
R_d	Gas constant of dry air
R_v	Gas constant of water vapor
R'	Effective gas constant for mixture of dry air and water vapor

t	Time
T	Temperature
T_d	Dew point temperature
T_{LCL}	Temperature of the Lifted Condensation Level
T_v	Virtual temperature
$T_{v,0}$	Standard virtual temperature
T'_v	Perturbation virtual temperature
u	x component of velocity (eastward)
v	y component of velocity (northward)
w	z component of wind velocity (upward)
x,y,z	Eastward, northward, and upward distance, respectively
α	Specific volume
κ	R_d/c_{pd}
ϵ	$= R_d/R_v$
ρ	Density
ρ_0	Standard density
ρ'	Perturbation density
θ	Potential temperature
θ_v	Virtual potential temperature
$\theta_{v,0}$	standard virtual potential temperature
θ'_v	deviation from the standard virtual potential temperature
θ_e	Equivalent potential temperature
θ_{ep}	Pseudoequivalent potential temperature
θ_w	Wet bulb potential temperature
ϕ	Latitude
Γ	$\equiv -dT/dz$, lapse rate of temperature
Γ_e	Environmental lapse rate
Γ_d	Dry abiabatic lapse rate
Γ_s	Pseudoadiabatic lapse rate
Ω	Angular speed of rotation of the earth or angular velocity of the earth.

Acronyms

ADC	Area of Deep Convection
AGL	Above Ground Level
AMF	Atmospheric Radiation Measurement Program Mobile Facility
AMSL	Above Mean Sea Level
ARM	Atmospheric Radiation Measurement Program
AWS	Automatic Weather Station
B	Buoyancy acceleration
BRN	Bulk Richardson number
CAA	Cold-Air Advection
CAP	Capping inversion
CAPE	Convective Available Potential Energy
Cb	Cumulonimbus cloud
CBL	Convective Boundary Layer
CCL	Convective Condensation Level
CIN	Convective Inhibition
COPS	Convective and Orographically-Induced Precipitation Study
CSIP	Convective Storm Initiation Project
DLR	Deutschen Zentrum für Luft- und Raumfahrt
DMC	Deep Moist Convection
DWD	Deutscher Wetterdienst
EL	Equilibrium Level
GFS	Global Forecast System (Model)
GFZ	GeoForschungsZentrum Potsdam
gpdm	geopotential decameters
GPS	Global Positioning System

HATPRO	Scanning Humidity and Temperature Profiler
IMK	Institute für Meteorologie und Klimaforschung
IOP	Intensive Operational Period
IR	Infrared radiation
ISDM	Inverse Square Distance Method
IWV	Integrated Water Vapor
LCL	Lifted Condensation Level
LFC	Level of Free Convection
LI	Lifted Index
LSI	Lid Strenght Index
MCS	Mesoscale Convective Systems
ML	Mixed Layer
MSG	Meteosat Secondary Generation
MU	Most Unstable (parcel)
NPW	Numerical Weather Prediction model
PBL	Planetary Boundary Layer
QG	Quasigeostrophic
RAD	Radiosonde
RS	Rapid Scanning
SB	Surface Based (parcel)
TSI	Total Sky Imager
UK	United Kingdom
VLBI	Very Long Baseline Interferometry
WAA	Warm-Air Advection
WVR	Water Vapor Radiometer

Bibliography

Andersson, T. A. M. J. C. N. S., 1989: Thermodynamic indices for forecasting thunderstorms in southern Sweden. *Meteorol. Mag.*, **116**, 141–146.

Anthes, R. A., 1984: Enhancement of convective precipitation by mesoscale variations in vegetative covering in semiarid regions. *J. Clim. Appl. Meteorol.*, **23**, 541–554.

Aoshima, F., A. Behrendt, H. S. Bauer, and V. Wulfmeyer, 2008: Statistics of convection initiation by use of Meteosat Rapid Scanning data during the Convective and Orographically-Induced Precipitation Study (COPS). *Submitted to Meteor. Zeitschr.*.

Banta, R. M., 1990: The role of mountain flows in making clouds. *Atmospheric Processes over Complex Terrain, Meteor. Monogr., Amer. Meteor. Soc.*, **45**, 229–283.

Barnes, S. L., 1964: A technique for maximizing details in numerical weather map analysis. *J. Appl. Meteor.*, **3**, 396–409.

Bastin, S., C. Champollion, O. Bock, P. Drobinski, and F. Masson, 2007: Diurnal Cycle of Water Vapor as Documented by a Dense GPS Network in a Coastal Area during ESCOMPTE IOP2. *Amer. Meteor. Soc.*, **46**, 167–182.

Bennett, L. J., K. A. Browing, A. M. Blyth, D. J. Parker, and P. A. Clark, 2006: A review of the initiation of precipitating convection in the United Kingdom. *Q. J. R. Meteorol. Soc.*, **132**, 1001–1020.

Bevis, M., S. Businger, T. Herring, C. Rocken, R. A. Anthes, and R. H. Ware, 1992: GPS Meteorology: Remote sensing of the atmospheric water vapor using the Global Positioning System. *J. Geophys. Res.*, **97**, 15787–15801.

Bluestein, H. B., 1993: *Synoptic-Dynamic Meteorology in Midlatitudes. Vol II.* Oxford Press, 594 pp.

Bock, O., E. Doerflinger, F. Masson, J. Walpersdorf, A. Van Baelen, J. Tarniewicz, M. Troller, A. Somieski, A. Geiger, and B. Buerki, 2004: GPS waper vapor project associated to the ESCOMPTE programme: description and first results of the field experiment. *Phys. Chem. Earth.*, **29**, 149–157.

Bolton, D., 1980: The Computation of Equivalent Potential Temperature. *Mon. Wea. Rev.*, **108**, 1046–1053.

Brooks, H. E., C. A. Doswell III, and L. J. Wicker, 1993: STORM-TIPE: A forecasting experiment using a three-dimensional cloud model. *Wea. Forecasting*, **8**, 352–362.

Browning, K., A. Blyth, P. Clark, U. Corsmeier, C. Morcrette, J. Agner, D. Bamber, C. Barthlott, L. Bennett, K. Beswick, M. Bitter, K. Bozier, B. Brooks, C. Collier, C. Cook, F. Davies, B. Deny, M. Engelhardt, T. Feuerle, R. Forbes, C. Gaffard, M. Gray, R. Hanken, T. Hewison, R. Huckle, N. Kalthoff, S. Khodayar, M. Kholer, S. Kraut, M. Kunz, D. Ladd, J. Lenfant, J. Marsham, J. McGregor, J. Nicol, E. Norton, D. Parker, F. Perry, M. Ramatschi, H. Roberts, A. Russel, H. Schulz, E. Slack, G. Vauhan, J. Waight, R. Watson, A. Webb, A. Wieser, and K. Zinz, 2007: The Convective Storm Initiation Project. *Bull. Am. Meteorol. Soc.*, 1–17.

Businger, S., S. R. Chiswell, M. Bevis, J. Duan, R. A. Anthes, C. Rocken, R. H. Ware, M. Exner, T. Van Hove, and S. Solheim, 1996: The promise of GPS in Atmospheric Monitoring. *Bull. Am. Meteorol. Soc.*, **77**, 5–18.

Byers, H. R. and H. R. Rodebush, 1948: Causes of thunderstorm of the Florida peninsula. *J. Meteor.*, **5**, 275–280.

Cady-Pereira, K. E., M. W. Shephard, D. D. Turner, E. J. Mlawer, S. A. Clough, and T. J. Wagner, 2007: Improved daytime column integrated precipitable water vapour from Vaisala radiosonde humidity sensors. *J. Atmos. Oceanic. Technol.*, **25**, 876–883.

Carbone, R., 1982: A severe frontal rainband. Part I: Stormwide hydrodynamics structure. *J. Atmos. Sci.*, **39**, 2639–2654.

Carbone, R. E., J. W. Conway, N. A. Crook, and M. W. Moncrieff, 1990: The generation and propagation of a nocturnal squall line. Part I: Observations and implications for mesoscale predictability. *Mon. Wea. Rev.*, **118**, 26–49.

Carbone, R. E., J. W. Wilson, T. D. Keenan, and J. M. Hacker, 2000: Tropical islands convection in the absence of significant topography. Part I: life cycle of diurnally forced convection. *Mon. Wea. Rev.*, **128**, 3459–3480.

Carlson, T. N., R. A. Anthes, M. Schwartz, S. G. Benjamin, and D. G. Baldwin, 1980: Analysis and prediction of severe storms environments. *Bull. Amer. Meteor. Soc.*, **61**, 1018–1032.

Carlson, T. N., S. G. Benjamin, G. S. Forbes, and Y. H. Lin, 1983: Elevated mixed layers in the regional severe storm envionment: Conceptual model and case studies. *Mon. Wea. Rev.*, **111**, 1453–1474.

Carlson, T. N. and F. H. Ludlam, 1968: Conditions for the occurrence of severe local storms. *Tellus.*, **20**, 203–226.

Charba, J. P., 1979: Two-to-six hour severe local storm probabilities: An operational forecasting system. *Mon. Wea. Rev*, **107**, 268–282.

Colby, F. P. J., 1984: Convective inhibition as a predictor of convection during AVE-SESAME II. *Mon. Wea. Rev*, **112**, 2239–2252.

Cressman, G. P., 1959: An operational objective analysis system. *Mon. Wea. Rev.*, **87**, 364–374.

Crook, N. A., 1996: Sensitivity of moist convection forced by boundary layer processes to low-level thermodynamic fields. *Mon. Wea. Rev*, **124**, 1768–1785.

Crook, N. A., T. L. Clark, and M. W. Moncrieff, 1991: The Denver Cyclone. Part II: Interaction with the convective boundary layer. *J. Atmos. Sci*, **48**, 2109–2126.

Dabberdt, W. F. and T. W. Schlatter, 1996: Research opportunities from emerging atmospheric observing and modeling capabilities. *Bull. Amer. Meteor. Soc.*, **77**, 305–323.

Daley, R., 1991: *Atmospheric data analyses*. University Press, Cambridge, 475 pp.

Daly, C., R. P. Neilson, and D. L. Phillips, 1994: A statistical-topographic model for mapping climatological precipitation over mountainous terrain. *J. Appl. Meteor.*, **33**, 140–158.

Davis, J. L., T. A. Herring, I. I. Shapiro, A. E. Rogers, and G. Elgered, 1985: Geodesy by radio interferometry: Effects of atmospheric modeling errors on estimates of baseline length. *Radio Sci.*, **20**, 1593–1607.

Dorninger, M., S. Schneider, and R. Steinacker, 2008: On the interpolation of precipitation data over complex terrain. *Meteorol. Atmos. Phys.*, **101**, 175–189.

Doswell, C., 1982: *The operational meteorology of convective weather.*, NOAA Tech. Rep., NWS NSSFC-5.

Doswell, C. I., H. E. Brooks, and R. A. Maddox, 1996: Flash Flood Forecasting: An Ingredients-Based Methodology. *Wea. Forecasting*, **11**, 560–581.

Doswell III, C., 1987: The distinction between large-scale and mesoscale contribution to severe convection: A case study example. *Wea. Forecasting*, **2**, 3–16.

Doswell III, C. A. and E. N. Rasmussen, 1994: The Effect of Neglecting the Virtual Temperature Correction on Cape Calculations. *Wea. Forecasting (Notes and Correspondence)*, **9**, 625–629.

Duan, J., M. Bevis, P. Fang, Y. Bock, S. Chiswell, S. Businger, C. Rocken, F. Solheim, T. Van Hove, R. H. Ware, S. McClusky, T. A. Herring, and R. W. King, 1996: GPS meteorology: Direct estimation of the absolute value of precipitable water. *J. Appl. Meteor.*, **35**, 830–838.

Emanuel, K. A., 1994: Atmospheric Convection. *Oxford University Press*, 580pp.

Emardson, T. R. and H. J. P. Derks, 1999: On the relation between the wet delay and the integrated precipitable water vapour in the European atmosphere. *Meteor. Appl.*, **6**, 1–12.

Emori, S., 1998: The interaction of cumulus convection with soil moisture distribution: An idealized simulation. *J. Geophys. Res.*, **103**, 8873–8884.

Fankhauser, J. C., C. J. Biter, C. G. Mohr, and R. L. Vaughan, 1985: Objective analysis of constant altitude aircraft measurements in thunderstorm inflow regions. *J. Atmos. Oceanic Technol.*, **2**, 157–170.

Fawbush, E. J. and R. C. Miller, 1952: A mean sounding representative of the tornadic air mass environment. *Bull. Amer. Meteor. Soc.*, **33**, 303–307.

Frei, C. and C. Schaer, 1998: A precipitation climatology of the Alps from high-resolution rain-gauge observation. *Int. J. Clim.*, **18**, 873–900.

Fuelberg, H. E. and D. G. Biggar, 1994: The preconvective environment of summer thunderstorms over the Florida Panhandle. *Wea. Forecasting*, **9**, 316–326.

Galway, J., 1958: The lifted index as a predictor of latent instability. *Bull. Amer. Meteor. Soc.*, **37**, 528–529.

Gandin, L., 1965: *Objective analysis of meteorological fields*. Israeli program for scientific translations, 242 pp.

Golding, B. W., 1998: Nimrod: A system for generating automated very short range forecasts. *Meteor. Appl.*, **5**, 1Ű16.

Grayson, R. B. and G. Bloeschl, 2000: *Spatial patterns in catchment hydrology: observations and modelling.*. Cambridge University Press, Cambridge, UK, 404 pp.

Graziano, T. M. and T. N. Carlson, 1987: A statistical evalutaion of lid strength on deep convection. *Weather and Forecasting*, **2**, 127–139.

Guichard, F., D. Parsons, and E. Miller, 2000: Thermodynamic and radiative impact of the correction of sounding humidity bias in the tropics. *J. Climate*, **13**, 3611–3624.

Haklander, A. J. and A. Van Delden, 2003: Thunderstorm predictors and their forecast skill for the Netherlands. *Atmos. Res.*, **68**, 273–299.

Holton, J. R., 1992: An Introduction to Dynamical Meteorology. *Academic Press*, 319pp.

Houze, R. A. J., 1993: *Cloud Dynamics*. Academic Press, 573 pp.

Huntrieser, H., H. Schiesser, W. Schimd, and A. Waldvolg, 1997: Comparison of traditional and newly developed thunderstorm indices for Switzerland. *Wea. Forecasting*, **12**, 108–125.

Hutchinson, M. F., 1995: Interpolating mean rainfall using thin plate smoothing splines. *Int. JGIS*, **9(4)**, 385–403.

John, R. H. and C. A. Doswell III, 1992: Severe local storms forecasting. *Wea. Forecasting*, **7**, 588–612.

Jorgensen, D. P. and T. M. Weckwerth, 2003: Forcing and organization of convective systems. Radar and Atmospheric Science: A Collection of Essays in Honor of Davis Atlas, R. M. Wakimoto and R. C. Srivastava. *Eds. Amer. Meteor. Soc.*, 75–104.

Kalthoff, N., B. Adler, C. Barthlott, U. Corsmeier, S. Mobbs, S. Crewell, K. Träumner, C. Kottmeier, A. Wieser, and V. Smith, 2009: The impact of convergence zones on the initiation of deep convection: A case study from COPS. *Atmos. Res. (accepted)*.

Khodayar, S., N. Kalthoff, M. Fiebig-Wittmaack, and M. Kohler, 2008: Evolution of the atmospheric boundary-layer structure of an arid Andes Valley. *Meteorol. Atmos. Phys.*, **99**, 181–198.

King, P. and M. L. Sills, 1998: The 1987 ELBOW Project: An experiment to study the effects of lake breezes on weather in southern ontario. *Preprints, 19th Conf. on Severe Local Storms, Mineapolis, MN, Amer. Meteor. Soc.*, 317–320.

Kingsmill, D. E., 1995: Convection initiation associated with a seabreeze front, a gust front, and their collision. *Mon. Wea. Rev.*, **123**, 2913–2933.

Kneifel, S., S. Crewell, U. Lohnert, and J. Schween, 2008: Investigating water vapor variability by ground-based microwave radiometry: evaluation using airborne observations. *IEEE Geoscience and Remote Sensing Letters*, 1–5.

Kocin, P. J., L. W. Uccellini, and R. A. Petersen, 1986: Rapid evolution of a jet streak circulation in a pre-convective environment. *Meteor. Atmos. Phys.*, **35**, 103–138.

Kottmeier, C., N. Kalthoff, C. Barthlott, U. Corsmeier, J. Van Baelen, A. Behrendt, R. Behrendt, A. Blyth, C. R., S. Crewell, P. Di Girolamo, M. Dorninger, C. Flamant, T. Foken, M. Hagen, C. Hauck, H. Hoeller, H. Konow, M. Kunz, H. Mahlke, S. Mobbs, E. Richard, R. Steineker, T. Weckwerth, A. Wieser, and V. Wulfmeyer, 2008: Mechanisms initiating deep convection over complex terrain during COPS. *Meteor. Zeitschr.*, **17**, 931–948.

Kunz, M., 2007: The skill of convective parameters and indices to predict isolated and severe thunderstorms. *Nat. Hazards Earth Syst. Sci.*, **7**, 327–342.

Laird, N. F., D. A. R. Kristovich, R. M. Rauber, H. T. Ochs, and L. J. Miller, 1995: The Cape Canaveral sea and river breezes: Kinematic structure and convective initiation. *Mon. Wea. Rev.*, **123**, 2942–2956.

Lauscaux, F., E. Richard, C. Keil, and O. Bock, 2004: Impact of the MAP reanalysis on the numerical simulation of the MAP-IOP2a convective system. *Meteor. Zeitschr.*, **13**, 49–54.

Lee, R. and J. E. Passner, 1993: The development and verification of TIPS: An expert system to forecast thunderstorm occurrence. *Wea. Forecasting*, **8**, 271–280.

Linne, H., B. Hennemuth, J. Bösenberg, and K. Ertel, 2007: Water vapour flux profiles in the convective boundary layer. *Theoretical and Applied Climatology*, **87**, 201–211(11).

Lorenc, A. C., D. Barker, R. S. Bell, B. Macpherson, and A. J. Maycock, 1996: On the use of Radiosonde Humidity Observations in mid-latitude NWP. *Meteorol. Atmos. Phys.*, **60**, 3–17.

Lucas, C. and E. J. Zipser, 2000: Environmental variability during TOGA COARE. *J. Atmos. Sci.*, **57**, 2333–2350.

Ludlam, F., 1963: Severe local storms: A review. *Severe Local Storms, Meteor. Monogr., No.27, Amer. Meteor. Soc.*, 1–30.

Maddox, R. A., F. Canova, and L. R. Hoxit, 1980: Meteorogical characteristics of flash flood events over the western United States. *Mon. Wea. Rev.*, **108**, 1866–1877.

Mahrt, L., 1976: Mixed layer moisture structure. *Mon. Wea. Rev.*, **104**, 1403–1407.

Manzato, A., 2005: The use of sounding-derived indices for neural network based short-term thunderstorm and rainfall forecasts. *Wea. Forecasting*, **20**, 896–917.

Marsham, J. H., A. M. Blyth, D. J. Parker, K. Beswick, K. A. Browning, U. Corsmeier, N. Kalthoff, S. Khodayar, C. J. Morcette, and E. G. Norton, 2007a: Variable cirrus shading during CSIP IOP5. II: Effects on the convective boundary layer. *Quart. J. Roy. Meteor. Soc.*, **133**, 1661–1675.

Marsham, J. H., C. J. Morcrette, K. A. Browing, A. M. Blyth, D. J. Parker, U. Corsmeier, N. Kalthoff, and M. Kholer, 2007b: Variable cirrus shading during CSIP IOP 5. I: Effects on the initiation of convection. *Quart. J. Roy. Meteor. Soc.*, **133**, 1643–1660.

Martin, L., M. Schneebeli, and C. Matzler, 2006: Tropospheric water and temperature retrieval for ASMUWARA. *Meteor. Zeitschr.*, **15**, 37–44.

McGinley, J. A., 1986: Nowcasting mesoscale phenomena. *Mesoscale Meteorology and Forecasting, P. S. Ray. Ed., Amer. Meteor. Soc.*, 657–688.

Meissner, C., N. Kalthoff, M. Kunz, and G. Adrian, 2007: Initiation of shallow convection in the Black Forest mountains. *Atmos. Res.*, **86**, 42–60.

Miller, R. C., 1975: *Notes on analysis and severe-storm forecasting procedures of the air force global weather contral*, AWS Tech. Rep. Vol. 200.

Moncrieff, M. W. and M. J. Miller, 1976: The dynamics and simulation of tropical cumulonimbus and squall lines. *Quart. J. Roy. Meteor. Soc.*, **102**, 373Ű394.

Morcrette, C., K. Browing, A. Blyth, K. Bozier, P. Clark, D. Ladd, E. Norton, and E. Pavelin, 2006: Secondary initiation of multiple bands of cumulonimbus over southern Britain. Part I: An observational case study. *Q. J. R. Meteorol. Soc.*, **132**, 1021–1051.

Newton, C., 1963: Dynamics of severe convective storms. Severe Convective Storms. *Meteor. Monogr. No.27, Amer. Meteor. Soc*, 33–58.

Normand, C. W. B., 1931a: Graphical indication of humidity in the upper air. *Nature*, **128**, 583.

— 1931b: Recent investigations on structure and movement of tropical storms in Indian Seas. *Gerlans Beitr.*, **34**, 233–243.

Ogura, Y. and N. A. Phillips, 1962: Scale analysis of deep and shallow convection in the atmosphere. *J. Atmos. Sci.*, **19**, 173–179.

Perry, K. D. and P. V. Hobbs, 1996: Influences of isolated cumulus clouds on the humidity of their surroundings. *J. Atmos. Sci.*, **53**, 159–174.

Pielke, R., 1974: A three-dimensional numerical model of the sea breezes over south Florida. *Mon. Wea. Rev.*, **102**, 115–139.

— 2001: Influence of the spatial distribution of vegetation and soils on the prediction of cumulus convective rainfall. *Reviews of Geophysics*, **39**, 151–177.

Pielke, R. A., G. Dalu, J. S. Snook, T. J. Lee, and T. G. F. Kittel, 1991a: Nonlinear influence of mesoscale land use on weather and climate. *J. Clim.*, **4**, 1053–1069.

Pielke, R. A. and M. Segal, 1986: Mesoscale circulations forced by differential terrain heating. Mesoscale Meteorology and Forecasting. *AMS, Boston, Chap. 22.*

Pielke, R. A., A. Song, P. J. Michaels, W. A. Lyons, and R. W. Arrit, 1991b: The predictability of sea-breeze generated thunderstorms. *Atmosfera*, **4**, 65–78.

Purdom, J. F. W., 1982: Subjetive interpretation of geostationary satellite data for nowcasting. *Nowcasting, K.A. Browning, Ed., Academic Press*, 149–166.

Rochette, S. M., J. T. Moore, and P. S. Market, 1999: The importance of parcel choice in elevated CAPE computations. *Natl. Wea. Dig.*, **23**, 20–32.

Rocken, C., R. Ware, T. VanHove, F. Solheim, C. Alber, J. Johnson, M. Bevis, and S. Businger, 1993: Sensing atmospheric water vapor with the Global Positioning System. *Geophys. Res. Lett.*, **20**, 2631–2634.

Rose, T. S. C., U. Lohnert, and C. Simmer, 2005: A network suitable microwave radiometer for operational monitoring of the cloudy atmosphere. *Atmos. Res.*, **75**, 183–200.

Saastamoinen, J., 1972: *Atmospheric correction for the troposphere and stratosphere in radio ranging of satellites*, In The Use of Artificial Satellites for Geodesy, Gheophys. Monogr. Ser., 15, 247-251.

Schaedler, G., 1990: Triggering of atmospheric circulations by moisture inhomogeneities of the earth's surface. *Boundary-Layer Meteorol.*, **51**, 1–20.

Schultz, D. M., P. N. Schumacher, and C. A. Doswell III, 2000: The Intricacies of Instabilites. *Bull. Amer. Meteor. Soc*, **128**, 4143–4148.

Schulz, P., 1989: Relationships of several stability indices to convective weather events in northeast Colorado. *Wea. Forecasting*, **4**, 73–80.

Segal, M. and R. W. Arritt, 1992: Nonclassical mesoscale circulations caused by surface sensible heat-flux gradients. *Bull. Am. Meteorol. Soc.*, **73**, 1593–1604.

Segal, M., R. W. Arritt, C. Clark, R. Rabin, and J. M. Brown, 1995: Scaling evaluation of the effect of surface characteristics on potential for deep convection over uniform terrain. *Mon. Wea. Rev.*, **123**, 383–400.

Segal, M., J. R. Garrat, G. Kallos, and R. A. Pielke, 1989: The impact of wet soil and canopy temperatures on daytime boundary-layer growth. *J. Atmos. Sci.*, **46**, 3673–3684.

Shepard, D., 1968: A two-dimensional interpolation function for irregularly spaced data. *Proc. 23rd ACM Nat. Conf. Brandon/Systems Press, Princeton, NJ*, 517–524.

Sherwood, S. C., 2000: On moist instability. *Mon. Wea. Rev.*, **128**, 4139–4142.

Skok, G. and T. Vrhovec, 2006: Considerations for interpolating rain gauge precipitation onto a regular grid. *Meteor. Zeitschr.*, **15**, 545–550.

Soden, B. and J. Lanzante, 1996: An assessment of satellite and radiosonde climatologies of upper-tropospheric water vapor. *J. Climate*, **9**, 1235–1250.

Stensrud, D. J. and R. A. Maddox, 1988: Mesoscale convective systems in weakly forced large-scale environments. Part II: Generation of mesoscale initiation condition. *Mon. Wea. Rev.*, **122**, 2068–2083.

Stull, R. B., 1988: *An Introduction to boundary layer meteorology*. Kluwer Academic Publisher, Boston.

Taylor, C. M., D. J. Parker, and P. Harris, 2007: An observational case study of mesoscale atmospheric circulations induced by soil moisture. *Geophys. Res. Lett.*, **34**, L15801, DOI:10.1029/2007GL030572.

Thiessen, A. H., 1911: Precipitation averages for large areas. *Mon. Wea. Rev.*, **39**, 1082–1084.

Turner, D. D., B. M. Lesht, S. A. Clough, J. C. Liljegren, H. E. Revercomb, and D. C. Tobin, 2003: Dry bias and variability in Vaisala radiosonses: The ARM experience. *J. Atmos. Oceanic Technol.*, **20**, 117–132.

Wahba, G. and J. Wendelberger, 1980: Some new mathematical methods for variational objective analysis using splines and cross validation. *Mon. Wea. Rev.*, **108**, 1122–1143.

Wakimoto, R. M., 2001: Convectively driven high wind events. *In Severe Convective Storms. Ed. C. Doswell III, American Meteorological Society, Boston, USA.*, 255–298.

Wang, J., H. L. Cole, D. J. Carlson, E. R. Miller, K. Beierle, A. Paukkunen, and T. K. Laine, 2002: Corrections of humidity measurement errors from the Vaisala RS80 radiosonde: Application to TOGA COARE data. *J. Atmos. Oceanic Technol.*, **19**, 981–1002.

Ware, R., C. Rocken, F. Solheim, M. Exner, W. Schreiner, R. Anthes, D. Feng, B. Herman, M. Gorbunov, S. Sokolovskiy, K. Hardy, Y. Kuo, X. Zou, K. Trenberth, T. Meehan, W. Melbourne, and S. Businger, 1996: GPS sounding of the atmosphere from low earth orbit: Preliminary results. *Bull. Amer. Meteor. Soc.*, **77**, 19–40.

Weckwerth, T. M., 2000: The effect of small-scale moisture variability on thunderstorm initiation. *Mon. Wea. Rev.*, **128**, 4017–4030.

Weckwerth, T. M. and et al., 2004: An overview of the International H_2 Project (IHOP_2002) and some preliminary highlights. *Bull. Amer. Meteor. Soc.*, **85**, 253–277.

Weckwerth, T. M. and D. B. Parsons, 2006: A Rewiew of Convection Initiation and Motivation for IHOP_2002. *Mon. Wea. Rev.*, **134**, 5–21.

Weckwerth, T. M., J. W. Wilson, and R. M. Wakimoto, 1996: Thermodynamic variability within the convective boundary layer due to horizontal convective rolls. *Mon. Wea. Rev.*, **124**, 769–784.

Weckwerth, T. M., V. Wulfmeyer, R. M. Wakimoto, R. M. Hardesty, J. W. Willson, and R. M. Banta, 1999: NCAR-NOAA lower-tropospheric water vapor workshop. *Bull. Amer. Meteor. Soc.*, **80**, 2339–2357.

Weisman, M. L. and J. B. Klemp, 1986: Characteristics of isolated convective storms. *Mesoscale Meteorology and Forecasting, P. S. Ray, Ed., Amer. Meteor. Soc.*, 331–358.

Wickert, J. and G. Gendt, 2006: GPS based remote sensing of the Earth's atmosphere. *pro. met. Jahrg. Nr.3/4*, **32**, 176–184.

Wilson, J., N. Crook, C. Mueller, J. Sun, and M. Dixon, 1998: Nowcasting thunderstorms: A status report. *Bull. Amer. Meteor. Soc.*, **79**, 2079–2100.

Wilson, J., G. Foote, N. Crook, C. Fankhauser, J.C.and Wade, J. Tuttle, C. Mueller, and S. K. Krueger, 1992: The role of boundary-layer convergence zones and horizontal rolls in the initiation of thunderstorms: A case study. *Mon. Wea. Rev.*, **120**, 1785–1815.

Wilson, J. and D. Megenhardt, 1997: Thunderstorm initiation, organization and lifetime associated with Florida boundary layer convergences lines. *Mon. Wea. Rev.*, **125**, 1507–1525.

Wilson, J. and W. Schreiber, 1986: Initiation of convective storms at radar-observed boundary-layer convergence lines. *Mon. Wea. Rev.*, **114**, 2516–2536.

Wulfmeyer, V., 1998: Ground-based differential absoption lidar for water-vapor and temperature profiling: development and specifications of a high-performance laser transmitter. *Appl. Optics*, **37**, 3804–3824.

Wulfmeyer, V., A. Behrendt, H.-S. Bauer, C. Kottmeier, U. Corsmeier, G. C. A. Blyth, U. Schumann, M. Hagen, S. Crewell, P. D. Girolamo, C. Flamant, M. Miller, A. Montani, S. Mobbs, E. Richard, M. Rotach, M. Arpagaus, H. Russchenberg, P. Schlüssel, M. König, V. Gärtner, R. Steinacker, M. Dorninger, D. Turner, T. Weckwerth, A. Hense, and C. Simmer, 2008: The convective and orographically-induced precipitation study. *Bull. Amer. Meteor. Soc.*, **89**, 1477–1486.

Yan, X., V. Ducroq, P. Poli, M. Hakam, G. Jaubert, and A. Walpersdorf, 2009: Impact of GPS zenith delay assimilation on convective scale prediction of Mediterranean heavy rainfall. *J. Geophys. Res.*, **114**.

Zhong, S., J. J. M. Leone, and E. S. Takle, 1991: Interaction of the sea breeze with a river breeze in an area of complex coastal heating. *Bound.-Layer Meteor.*, **56**, 101–139.

Zieger, C. L., T. J. Lee, and R. A. Pielke Sr., 1997: Convection initiation at the dryline: A modeling study. *Mon. Wea. Rev.*, **125**, 1001–1026.

Acknowledgements

I would like to thank my supervisor Prof. Dr. C. Kottmeier for his support and availability during all my time in the IMK.

Many thanks to Prof. Dr. S. Jones for co-directing my thesis and for all her instructive comments on my work.

I would like especially to acknowledge Dr. N. Kalthoff for his trust, for all the helpful discussions and for sharing part of his knowledge with me. Thanks to my collegues from the "Land surfaces and Boundary layer" working group.

Many thanks to Pieter Groenemeijer, my roommate during these years, for all the interesting discussions.

Also I wish to thank all my collegues at the IMK and at the meteorology department of Karlsruhe University for the nice company during all this time.

I acknowledge the DFG-funded Graduiertenkolleg "Naturkatastrophen" program at Karlsruhe University for the financial support and all the Ph.D. students I met during this period.

Para mi familia, Paco, mi padre, mi madre, mi hermana y hermano. Esto es por y para vosotros. Nunca hubiese sido posible sin vuestro amor incondicional.

Wissenschaftliche Berichte des Instituts für Meteorologie und Klimaforschung des Karlsruher Instituts für Technologie (0179-5619)

Bisher erschienen:

Nr. 1: *Fiedler, F. / Prenosil, T.*
Das MESOKLIP-Experiment. (Mesoskaliges Klimaprogramm im Oberrheintal).
August 1980

Nr. 2: *Tangermann-Dlugi, G.*
Numerische Simulationen atmosphärischer Grenzschichtströmungen über langgestreckten mesoskaligen Hügelketten bei neutraler thermischer Schichtung.
August 1982

Nr. 3: *Witte, N.*
Ein numerisches Modell des Wärmehaushalts fließender Gewässer unter Berücksichtigung thermischer Eingriffe.
Dezember 1982

Nr. 4: *Fiedler, F. / Höschele, K. (Hrsg.)*
Prof. Dr. Max Diem zum 70. Geburtstag.
Februar 1983 (vergriffen)

Nr. 5: *Adrian, G.*
Ein Initialisierungsverfahren für numerische mesoskalige Strömungsmodelle.
Juli 1985

Nr. 6: *Dorwarth, G.*
Numerische Berechnung des Druckwiderstandes typischer Geländeformen.
Januar 1986

Nr. 7: *Vogel, B.; Adrian, G. / Fiedler, F.*
MESOKLIP-Analysen der meteorologischen Beobachtungen von mesoskaligen Phänomenen im Oberrheingraben.
November 1987

Nr. 8: *Hugelmann, C.-P.*
Differenzenverfahren zur Behandlung der Advektion.
Februar 1988

Nr. 9: *Hafner, T.*
Experimentelle Untersuchung zum Druckwiderstand der Alpen.
April 1988

Nr. 10: *Corsmeier, U.*
Analyse turbulenter Bewegungsvorgänge in der maritimen atmosphärischen Grenzschicht.
Mai 1988

Nr. 11: *Walk, O. / Wieringa, J.(eds)*
Tsumeb Studies of the Tropical Boundary-Layer Climate.
Juli 1988

Nr. 12: *Degrazia, G. A.*
Anwendung von Ähnlichkeitsverfahren auf die turbulente Diffusion in der konvektiven und stabilen Grenzschicht.
Januar 1989

Nr. 13: *Schädler, G.*
Numerische Simulationen zur Wechselwirkung zwischen Landoberflächen und atmophärischer Grenzschicht.
November 1990

Nr. 14: *Heldt, K.*
Untersuchungen zur Überströmung eines mikroskaligen Hindernisses in der Atmosphäre.
Juli 1991

Nr. 15: *Vogel, H.*
Verteilungen reaktiver Luftbeimengungen im Lee einer Stadt – Numerische Untersuchungen der relevanten Prozesse.
Juli 1991

Nr. 16: *Höschele, K.(ed.)*
Planning Applications of Urban and Building Climatology – Proceedings of the IFHP / CIB-Symposium Berlin, October 14-15, 1991.
März 1992

Nr. 17: *Frank, H. P.*
Grenzschichtstruktur in Fronten.
März 1992

Nr. 18: *Müller, A.*
Parallelisierung numerischer Verfahren zur Beschreibung von Ausbreitungs- und chemischen Umwandlungsprozessen in der atmosphärischen Grenzschicht.
Februar 1996

Nr. 19: *Lenz, C.-J.*
Energieumsetzungen an der Erdoberfläche in gegliedertem Gelände.
Juni 1996

Nr. 20: *Schwartz, A.*
Numerische Simulationen zur Massenbilanz chemisch reaktiver Substanzen im mesoskaligen Bereich.
November 1996

Nr. 21: *Beheng, K. D.*
Professor Dr. Franz Fiedler zum 60. Geburtstag.
Januar 1998

Nr. 22: *Niemann, V.*
Numerische Simulation turbulenter Scherströmungen mit einem Kaskadenmodell.
April 1998

Nr. 23: *Koßmann, M.*
Einfluß orographisch induzierter Transportprozesse auf die Struktur der atmosphärischen Grenzschicht und die Verteilung von Spurengasen.
April 1998

Nr. 24: *Baldauf, M.*
Die effektive Rauhigkeit über komplexem Gelände – Ein Störungstheoretischer Ansatz.
Juni 1998

Nr. 25: *Noppel, H.*
Untersuchung des vertikalen Wärmetransports durch die Hangwindzirkulation auf regionaler Skala.
Dezember 1999

Nr. 26: *Kuntze, K.*
Vertikaler Austausch und chemische Umwandlung von Spurenstoffen über topographisch gegliedertem Gelände.
Oktober 2001

Nr. 27: *Wilms-Grabe, W.*
Vierdimensionale Datenassimilation als Methode zur Kopplung zweier verschiedenskaliger meteorologischer Modellsysteme.
Oktober 2001

Nr. 28: *Grabe, F.*
Simulation der Wechselwirkung zwischen Atmosphäre, Vegetation und Erdoberfläche bei Verwendung unterschiedlicher Parametrisierungsansätze.
Januar 2002

Nr. 29: *Riemer, N.*
Numerische Simulationen zur Wirkung des Aerosols auf die troposphärische Chemie und die Sichtweite.
Mai 2002

Nr. 30: *Braun, F. J.*
Mesoskalige Modellierung der Bodenhydrologie.
Dezember 2002

Nr. 31: *Kunz, M.*
Simulation von Starkniederschlägen mit langer Andauer über Mittelgebirgen.
März 2003

Nr. 32: *Bäumer, D.*
Transport und chemische Umwandlung von Luftschadstoffen im Nahbereich von Autobahnen – numerische Simulationen.
Juni 2003

Nr. 33: *Barthlott, C.*
Kohärente Wirbelstrukturen in der atmosphärischen Grenzschicht.
Juni 2003

Nr. 34: *Wieser, A.*
Messung turbulenter Spurengasflüsse vom Flugzeug aus.
Januar 2005

Nr. 35: *Blahak, U.*
Analyse des Extinktionseffektes bei Niederschlagsmessungen mit einem C-Band Radar anhand von Simulation und Messung.
Februar 2005

Nr. 36: *Bertram, I.*
Bestimmung der Wasser- und Eismasse hochreichender konvektiver Wolken anhand von Radardaten, Modellergebnissen und konzeptioneller Betrachtungen.
Mai 2005

Nr. 37: *Schmoeckel, J.*
Orographischer Einfluss auf die Strömung abgeleitet aus Sturmschäden im Schwarzwald während des Orkans „Lothar".
Mai 2006

Nr. 38: *Schmitt, C.*
Interannual Variability in Antarctic Sea Ice Motion: Interannuelle Variabilität antarktischer Meereis-Drift.
Mai 2006

Nr. 39: *Hasel, M.*
Strukturmerkmale und Modelldarstellung der Konvektion über Mittelgebirgen.
Juli 2006

Ab Band 40 erscheinen die Wissenschaftlichen Berichte des Instituts für Meteorologie und Klimaforschung bei KIT Scientific Publishing (ISSN 0179-5619).
Die Bände sind unter www.ksp.kit.edu als PDF frei verfügbar oder als Druckausgabe bestellbar.

Nr. 40: *Lux, R.*
Modellsimulationen zur Strömungsverstärkung von orographischen Grundstrukturen bei Sturmsituationen. (2007)
ISBN 978-3-86644-140-8

Nr. 41: *Straub, W.*
Der Einfluss von Gebirgswellen auf die Initiierung und Entwicklung konvektiver Wolken. (2008)
ISBN 978-3-86644-226-9

Nr. 42: *Meißner, C.*
High-resolution sensitivity studies with the regional climate model COSMO-CLM. (2008)
ISBN 978-3-86644-228-3

Nr. 43: *Höpfner, M.*
Charakterisierung polarer stratosphärischer Wolken mittels hochauflösender Infrarotspektroskopie. (2008)
ISBN 978-3-86644-294-8

Nr. 44: *Rings, J.*
Monitoring the water content evolution of dikes. (2009)
ISBN 978-3-86644-321-1

Nr. 45: *Riemer, M.*
Außertropische Umwandlung tropischer Wirbelstürme: Einfluss auf das Strömungsmuster in den mittleren Breiten. (2012)
ISBN 978-3-86644-766-0

Nr. 46: *Anwender, D.*
Extratropical Transition in the Ensemble Prediction System of the ECMWF: Case Studies and Experiments. (2012)
ISBN 978-3-86644-767-7

Nr. 47: *Rinke, R.*
Parametrisierung des Auswaschens von Aerosolpartikeln durch Niederschlag. (2012)
ISBN 978-3-86644-768-4

Nr. 48: *Stanelle, T.*
Wechselwirkungen von Mineralstaubpartikeln mit thermodynamischen und dynamischen Prozessen in der Atmosphäre über Westafrika. (2012)
ISBN 978-3-86644-769-1

Nr. 49: *Peters, T.*
Ableitung einer Beziehung zwischen der Radarreflektivität, der Niederschlagsrate und weiteren aus Radardaten abgeleiteten Parametern unter Verwendung von Methoden der multivariaten Statistik. (2012)
ISBN 978-3-86644-323-5

Nr. 50: *Khodayar Pardo, S.*
High-resolution analysis of the initiation of deep convection forced by boundary-layer processes. (2015)
ISBN 978-3-86644-770-7